Alfred Elwes

How I crossed Africa

From the Atlantic to the Indian Ocean. Vol. 1: The King´s Rifle

Alfred Elwes

How I crossed Africa
From the Atlantic to the Indian Ocean. Vol. 1: The King´s Rifle

ISBN/EAN: 9783337257255

Printed in Europe, USA, Canada, Australia, Japan

Cover: Foto ©Andreas Hilbeck / pixelio.de

More available books at **www.hansebooks.com**

HOW I CROSSED AFRICA:

FROM THE

ATLANTIC TO THE INDIAN OCEAN, THROUGH UNKNOWN COUNTRIES;
DISCOVERY OF THE GREAT ZAMBESI AFFLUENTS, &c.

By MAJOR SERPA PINTO.

TRANSLATED FROM THE AUTHOR'S MANUSCRIPT
By ALFRED ELWES.

IN TWO VOLUMES.
CONTAINING 15 MAPS AND FACSIMILES, AND 132 ILLUSTRATIONS.

Vol. I.—THE KING'S RIFLE.

LONDON:
SAMPSON LOW, MARSTON, SEARLE, & RIVINGTON,
CROWN BUILDINGS, 188 FLEET STREET.
1881.

[All rights reserved.]

TO HIS MAJESTY THE KING,

D. LUIZ I.,

BY GRACIOUS PERMISSION

THIS BOOK IS DEDICATED BY

THE AUTHOR.

SIRE,

It was no feeling of servile adulation which induced me to pray Your Majesty's permission to dedicate to you this Book; it was rather the recognition of a double debt of justice and gratitude,—of justice to the intelligent and enlightened monarch who signed the decree which created resources for the first Portuguese scientific expedition of the present century to Central Africa,—of gratitude to the prince whose endowments, both of heart and mind, are on a par with his lofty qualities, and render him one of the first constitutional rulers of contemporary Europe.

Your Majesty gave me the opportunity of connecting the obscure name of a Portuguese soldier with one of the happiest and most auspicious attempts essayed in modern times by Portugal.

And this work belongs therefore to Your Majesty, as a legitimate title of my profound gratitude. I consequently venture respectfully to entreat Your Majesty to be good enough to accept the humble offering in the same benevolent spirit wherewith you deigned to spur me on to an enterprise of which, at its close, Your Majesty's favours were still held to be the sincerest and most treasured recompense.

    Your aide-de-camp
        And most devoted of
                Your subjects,
                      ALEXANDRE DE SERPA PINTO.

LONDON, 61 GOWER STREET,
   5th December, 1880.

TO HIS EXCELLENCY, COUNSELLOR JOÃO D'ANDRADE CORVO.

Most Illustrious and Excellent Sir,

In submitting my name, in 1877, to the Central Permanent Commission of Geography with a view to my forming part of the Portuguese expedition to the interior of Africa, Your Excellency assumed the responsibility of my nomination.

It was my constant desire to give Your Excellency the fullest satisfaction for thus venturing to select me for the performance of so arduous a task.

This book contains, together with a narrative of my adventures, the results of my labours and studies.

I do not know whether they will come up to Your Excellency's expectations; and I am just as ignorant whether I have properly fulfilled the duties which Your Excellency, in the name of our country, intrusted to me.

I have, however, the consciousness that I did my best, and that I followed out, so far as human strength enabled me to do, Your Excellency's ideas and instructions.

A perusal of my narrative will show Your Excellency with how many difficulties I struggled, and how poor were the resources I at last had at my disposal.

If, however, the results of my labours are found to correspond to the confidence with which Your Excellency was good enough to honour me, they will constitute the highest praise to which I can aspire, being, as I am, the most respectful admirer of Your Excellency's talent, vast knowledge, and elevated qualities.

ALEXANDRE DE SERPA PINTO.

A TRIBUTE OF GRATITUDE.

I AM about to cite names. It is a difficult and a dangerous task. There is always a fear of wounding the modest or hurting the susceptible. Nevertheless I must venture.

The list will be a long one, as the favours were many in number; and I may perchance sin in the way of omission, the offspring of a slothful memory.

May I obtain pardon, both from those who would wish to hide their kindness behind the veil of modesty, and from those whom a slip of remembrance may have caused me to leave unnamed.

Following the chronological order of facts, I will endeavour to recall, with a deep feeling of gratitude, the many services and kindnesses I have received.

To the Central Geographical Commission belongs the first place in my estimation, for having selected me as its instrument in the work of exploration which it had decided upon making in Africa.

Proposed by H.E. Counsellor Andrade Corvo, I was unanimously accepted by that learned body, and the suggestions which I presented for the organisation of the undertaking were duly attended to. While speaking of the Central Geographical Commission, I cannot refrain from mentioning individuals, for, though I received courtesy from all, I was especially assisted by many.

Dr. Bernardino Antonio Gomes, the Marquis de Sousa-Hollstein, Antonio Augusto Teixera de Vasconcellos, are names which ought not to escape a record of my gratitude, though their owners are now lying in their graves.

Dr. Julio Rodriguez, Luciano Cordeiro, Dr. Bocage, Count de Ficalho, Carlos Testa, Pereira da Silva, Jorge Figaniere and Francisco da Costa e Silva, were the gentlemen who, from their position at the Board, most endeavoured to honour me with their favour.

Another with whom I only became personally acquainted years afterwards, who was absent while the expedition was being organised, lent his valuable counsel in respect of the scientific portions of the enterprise. I refer to Mr. Brito Limpo.

Outside of the Society, I obtained valuable assistance from my intimate friends Murrecas Ferreira and João Botto.

After the Central Commission comes the Geographical Society of Lisbon, and prominently among its members are its Presidents, Dr. Bocage and Viscount de St. Januario, and its Secretaries Luciano Cordeiro and Rodrigo Pequito.

The Portuguese Journals follow in a natural course, and I cordially thank their editors for all the favours they have conferred upon me, and for the manner in which they hailed my appointment.

Beyond my own country most valuable aid was afforded me by Mendes Leal, Antonio d'Abbadie and Ferdinand de Lesseps in Paris; and Viscount de Duprat and Lieutenant Pinto da Fonseca Vaz in London; inasmuch as the co-operation of those gentlemen, and that only, enabled Capello and myself to carry out our resolve, to have all the material of the expedition organised within a month.

Before quitting Portugal, two other gentlemen must

be remembered, as they had much to do with the ultimate success of our enterprise.

These are Counsellor José de Mello e Gouvea, who was then intrusted with the portfolio of Ultramar, and Francisco Costa, the Director-General of the Ministry of the Colonies.

Pedro d'Almeida Tito and Avelino Fernandes showed me so much attention on my journey, that I cannot refrain from mentioning them here.

After them follow the Governor of Cape Verde, Vasco Guedes, and the Governor of Angola, Caetano d'Albuquerque; both of whom displayed the utmost kindness.

In Loanda, José Maria de Prado, Urbano de Castro, Consul Newton, the *Associação Commercial* and, above all, the officers and commander of the gunboat *Tamega* earned a title to my deepest gratitude.

And now comes a name which at that time was being echoed from every part of the globe, to the remotest corners of which it had penetrated—

Henry Moreland Stanley.

The great explorer, the intrepid traveller, who had just terminated the most stupendous journey of modern times, was my friend and my counsellor, from whom I received the most valuable lessons. A better master could not well be obtained. I will beg him to accept, in these brief lines, the sincerest tribute of the great admiration I feel for him, and the frankest expression of my esteem, and of the gratitude with which he has inspired me.

In Benguella, Pereira de Mello and Silva Porto occupy the first place: I need not stay to speak of them at greater length, as their acts, narrated in these volumes, constitute their highest praise. Antonio Ferreira Marques, Lieutenant Seraphim, the chemist Monteiro, and Vieira da Silva, are names which I cannot pass unnoticed.

Santos Reis, my host of the Dombe Grande, and Lieutenant Roza of Quillengues, are two more creditors to my gratitude.

I will now make a prodigious leap, and without stopping at Dr. Bradshaw and the Coillard family, transport myself to Bamanguato, to Shoshong, where the favours of King Kama, and above all those of Mr. and Mrs. Taylor will ever dwell in my memory.

But no light difficulty next presents itself. I am in Pretoria—in the first civilised portion of the world I fell in with after leaving Benguella; and where the favours heaped upon me were so many that I know not how to find words or space to record the names of their bestowers.

Mr. Swart, the Government Treasurer, was the first to honour me, and he is of right the first to be recorded.

After him come Frederick Jeppe, Secretary Osborne, Dr. Rissick, Mr. Kisch, Major Tylor and Captain Saunders, together with all the officers of the 80th regiment.

Baroness van Levetzow, Mrs. Imink and Mrs. Kisch, and finally Colonel Lanyan.

Sir Bartle Frere immediately came to my aid, nor was our Portuguese Consul at the Cape, Snr. Carvalho, far behind.

If I owe a debt of gratitude to the English Governor, I am no less beholden to the Portuguese Consul, who, by means of immediate telegrams, hastened to render me the utmost assistance.

Monseigneur Jolivet, the learned Bishop of Natal, then residing at Pretoria, was not among the last to load me with favours.

On my way to Durban, I received the utmost courtesy from Mr. Goodliffe, and when at Maritzburg those courtesies were repeated by Colonel Baker, Captain Whalley, Mrs. Saunders, and Mr. Furse.

In Durban, Mr. Snell, the Portuguese Consul, and Mr. and Mrs. B. H. de Waal, the gentleman at the head of the Handels Company of Eastern Africa, were foremost in their kindness and attention.

The task I have imposed upon myself becomes more and more embarrassing. I am on my way to Europe after the completion of my journey, and favours and courtesy meet me at every stage.

In Lourenço Marquez I have to mention Castilho, Machado, Maia, and Fonseca. In Mozambique, Governor Cunha, Torrezão, and in fact every one.

In Zanzibar, Dr. and Mrs. Kirk, Widmar, and above all Captain Draper of the *Danubio*, belonging to the Union Steam Ship Company, who conveyed me thither from Durban, should not remain unnamed.

In Cairo, again, Widmar was most kind. In Alexandria, Count and Countess de Caprara especially deserved my acknowledgments.

Even before I reached Lisbon, I received an important service from the Baron de Mendonça at Bordeaux.

In Lisbon, the Government in the first instance, and friends old and new, vied with each other in showing me attention.

I remained there some ten days, which were all too short to acknowledge the acts of courtesy shown me, and which left me not a minute for thanks.

Barely recovered from the fatigues of my voyage, it was expected that I should deliver a lecture upon my wanderings, and without the valuable assistance rendered me by Pequito, Sarrea Prado, Batalha Reis, and Dr. Bocage, it would have been impossible for me to pull through.

Not wishing or rather not being able here to mention other names, so great is their number, I must content myself with thanking, in the warmest terms of gratitude, the Geographical Society of Lisbon, for all their attentions towards me.

To the *Associação Commercial*, and to its worthy President Snr. Chamisso, who all through displayed the utmost interest in the exploration of which I formed part, I tender my heartfelt acknowledgments.

I learned when at Lisbon a fact which I cannot refrain from recording here, and coupling with it a name, that of Snr. Thomaz Ribeiro. To him I am indebted for the orders he gave, in his capacity of Minister of Marine, that assistance should be sent me from Mozambique into the interior of Africa, and I herewith tender him my grateful thanks.

I beg also to express my deep acknowledgments to the Diplomatic Corps resident in Lisbon, and above all to M. Morier, Baron de P. Hegeurt, Laboulay, the Marquis d'Oldoini and Ruata.

My thanks are also due to the *Associação Commercial* of Oporto, to the voluntary Fire Brigade of that city, to the *Sociedade Euterpe*, and the *Sociedade de Instrucção*, as well as to the municipalities and other institutions of the country, who did me honour.

To the Portuguese Associations in Brazil, to my fellow-countrymen, who were so kind to me when far from home, to all those who spared neither time nor pains on my behalf, I waft friendly greetings and expressions of my gratitude.

Above all are they due to those who formed a Society bearing my name, and from distant Pernambuco offered me this delicate compliment, which I am not likely ever to forget.

It is now my pleasing duty in the order of events to offer my thanks to foreign Sovereigns for the high honours which they conferred upon me, more especially to His Majesty of Belgium, the illustrious and learned King Leopold, the great instigator of modern geographical discovery in Africa, who, apart from the honour with which he deigned to distinguish me, gave me marks of the most cordial esteem, and expressed the kindest interest in my welfare.

The Geographical Societies of France, and chiefly those of Paris, in the persons of Admiral La Roncière le Noury, Ferdinand de Lesseps, Messrs. Daubré, Maunoir, d'Abbadie, de Quatrefages and Duveyrier, were profuse in their favours; the Society at Marseilles, conferred upon me a lofty and cherished distinction, and its President, M. Babaud, showed me the utmost courtesy; nor must I forget the kindness of the *Société Commerciale* of Paris, or of its worthy Secretary-General, M. Gauthiot.

Referring to Paris, I cannot leave unnoticed the Portuguese Colony, nor among its members, the names of Mendes Leal, Count de S. Miguel, Camillo de Moraes, Pereira Leite, Garrido and Dr. Aguiar, whose friendly attentions will never be effaced from my mind.

I also feel deeply indebted to the Belgian Geographical Societies, and particularly to that of Anvers, in the persons of their Presidents, General Liagre and Colonel Wauvermans; nor should I fail to record, in a country where I met with universal courtesy, the names of Messrs. du Fief, Bamps and Colonel Strauch, and above all that of Count de Thomar, whose repeated favours and cordiality of treatment converted the sincere esteem of our first relations into what I trust will be a lasting friendship.

In the order of dates I now come to England, the last upon my list, but perhaps the first in point of importance, owing to the numerous acts of courtesy and recognition which I there met with.

My gratitude was first awakened in the English Colonies of South Africa, and it was increased tenfold by my reception in the mother-country.

I cannot possibly name all those to whom I here proffer my most grateful thanks, but I would specially express my acknowledgments to the Geographical Society of London, to its President the Earl of Northbrook, to its Secretaries Clements R. Markham and Bates, and to its members Sir Rutherford Alcock, Lord Arthur Russell, and Viscount de Duprat.

To Mr. Frederick Youle, Dr. Peacock and Messrs. M. d'Antas, Sampaio, Fonseca Vaz, Quillinan, Duprat and Ribeiro Saraiva, I owe, besides the most cordial attentions, a deep debt of gratitude for services rendered during my serious illness, for which I offer them here a public acknowledgment.

Ere I conclude, I must not omit to mention the names of Mr. David Ward, the Mayor of Sheffield, and of my particular friend the great and eminent explorer, Verney Lovett Cameron, and with them I must close a list which is likely otherwise to be interminable.

To the Scientific Societies of other countries and to all those who are not otherwise alluded to, but from whom I received distinguished favours, I beg to express my warmest acknowledgments, not the less sincere because they are not individually awarded.

MAJOR ALEXANDRE DE SERPA PINTO.

LONDON, *5th December*, 1880.

THE BOOK.

This book has no pretensions to a literary work.

Written without much attention to form, it is a faithful reproduction of my travelling diary.

I have eliminated from it many episodes of the chase, and other matters which may, during my intervals of leisure, constitute a volume of a special character. I have endeavoured, above all else, to put prominently forward that which I deemed most interesting in the way of geographic and ethnographic research; and if I have occasionally interwoven a few of the many dramatic episodes which abounded in my fatiguing enterprise, I have done so where they became connected with notable facts of sufficient importance to alter a projected itinerary, to determine my stay in, or my precipitate march from, any place, which would be incomprehensible without an explanation of the causes which led up to such resolve.

To a European, and generally to any man who has never travelled in the wilds of Africa, what explorers have to endure in penetrating into that continent, what difficulties they have at every instant to overcome, and what iron labour they have to go through, will be well nigh incomprehensible.

The narratives of Livingstone, Cameron, Stanley, Burton, Grant, Savorgnan de Brazza, d'Abbadie, Ed. Mohr, and many others, are far from depicting all the sufferings of the African traveller. It is difficult, in fact, for any one to conceive them who has not experienced them in his own person; and it is equally difficult for the man who has endured them to describe them properly.

I do not even attempt to portray what I suffered, or endeavour to show the amount of work I had to perform. Whether those who calmly examine the result of my labours will or will not give me the just meed which I consider rightly my due, is to-day a matter of indifference to me, being, as I am, convinced that I can only be properly judged by those who, like myself, have trodden the almost endless tracts of the Dark Continent, and have undergone the wretchedness and privations which were too often my lot in the country.

Just as only that man who, being a father, can comprehend the bitter grief occasioned by the loss of a beloved child, so only he who has been himself an explorer can thoroughly appreciate the tribulations that a brother explorer has to endure.

The facts narrated in this book are the expression of the truth; a bitter truth indeed at times, but which it would be a deep wrong to conceal.

I have endeavoured to present therein the results of a ceaseless labour of many months, and I vouch for what I record about African geography, as being the sole authority to speak upon the subject in what concerns my own special journey; and I shall continue to vouch for the correctness of my data, until some other man shall follow in my steps across the African continent, and convince me that I am in error.

The general opinions which I enunciate touching this or that problem may be erroneous; they are of course open to criticism, and may fall to the ground beneath the practical demonstration of future journeys, in the same manner as have fallen the assertions of many of my most illustrious predecessors; but what I hold to be incontestable and impossible to be contested, are the facts which I saw, and the data which refer to the countries I passed through, and which I describe in this book with the conscientiousness that ought ever to dictate the records of the explorer.

I did not repair to Africa with a view to gain money. I had but the scanty pay of an officer in the army, and I sought no other.

I left behind me a family that I held most dear. I left my country and all its attractions, for a weary labour, and for the sole purpose of labouring, in co-operation with other countries, in the great task of survey of the unknown continent; and I feel the consciousness of having done so with all my strength.

I leave to men of science, and to those who are authorities in such matters, to appreciate my work according to its deserts.

I say no more upon a subject which may, perhaps, appear to spring from a spirit of vanity to which I am a stranger, but circumstances of an unusual kind, which occurred during the early months of my residence in Europe, after completing my weary African journey, dictated the lines which I have above written.

A year has passed since I began to reduce to shape the results of my African labours, but an obstinate illness, again and again, stood in the way of my ardent desire to lay those labours before the public.

Commenced in London in September 1879, my book was almost entirely written in the months of September and October 1880, at the Figueira da Foz in Portugal.

The haste with which it was terminated will no doubt have contributed greatly to its imperfection of form.

It is published in London, where, with the eminent firm of Sampson Low, Marston, Searle and Rivington,

I met with facilities which I could scarcely have found elsewhere.

Those gentlemen did not hesitate to incur the enormous expense naturally inherent to so difficult and costly a publication, and they have been good enough to undertake to print in England the Portuguese edition; a most onerous task, where the difference between the two languages necessitated the founding of new type, owing to the characteristic marks of our southern idiom.

I am deeply indebted to them for the care and attention they have devoted to the work, to the merit of which, if it be held to possess any, they have certainly largely contributed.

The book was written in Portuguese, and its translation was kindly undertaken by Mr. Alfred Elwes, the well-known English writer and philologist. I beg to express my warmest acknowledgments to that gentleman, for the way in which he has interpreted my ideas and faithfully translated my phrases—a task of the greater difficulty, as the richness and intricacy of the Portuguese language are considerable. On perusing the English translation, I have again and again admired the closeness with which my style has been adhered to; for my phrases are laconic, and therefore all the more stubborn to deal with in a foreign language. If the book have any value, it has certainly lost nothing by translation into English, and to Mr. Elwes the honour of executing it so conscientiously is due.

Before closing these remarks, I wish also to thank most sincerely Counsellor Antonio Ribeiro Saraiva, who, notwithstanding his own duties and his advanced age, was good enough to do me the special favour to correct the Portuguese proofs; Mr. E. Weller, the cartographer, who undertook the engraving of my maps; and Mr. Cooper, who has so admirably succeeded in interpreting my hasty sketches, made during the journey, in the engravings which illustrate the work.

Here then is the book. My sole desire is that it may interest and entertain the mass of my readers, serve as a study to others, and give a fresh impulse to the great and sublime crusade of the nineteenth century, a crusade of civilisation and progress in the Dark Continent.

LONDON, 61 GOWER STREET,
 5th *December*, 1880.

THE TITLE OF THE BOOK.

On my return to my temporary home this evening, from an after-dinner stroll, I find upon my writing-table, pinned to the blotter, a cutting from some newspaper, which contains the following words:—

"The *Athenæum* says, 'Major Serpa Pinto, who has recovered from his protracted illness, has come to London to bring out his book, descriptive of his journey across Africa. It is satisfactory to find that the title is altered from "The King's Rifle," to "How I crossed Africa." "The King's Rifle" might be a good name for a boy's book of adventures by Mayne Reid or Gustave Aimard, but it seems rather out of place on the title-page of a serious book of African travels.'"

It is near midnight, and I feel that I want rest; but before turning in, I cannot refrain from writing a few words upon the above subject.

The critic's remark is, and yet it is not, quite correct.

African travels always partake, more or less, of romance, however much they may take the form of a scientific work.

If my book, like all which have preceded it, is a veritable romance, it nevertheless contains geographic matter of some importance.

The project which I formed, and which I have here carried out, was to blend with a narrative of my adventures the more serious labours referred to; just as such things are apt to be blended in the wilds of Africa.

As to the title that my book should bear, it was a matter to which I gave but little heed.

Inasmuch as the expedition, and, as a necessary consequence, the whole fruits of my labour were saved by the King's Rifle, it occurred to me to give that title to my entire work. I gave no thought to the adverse criticism it might meet with. And besides, my justification would be found in a perusal of my narrative.

One consideration, nevertheless, occurred to modify my original project.

One man there was, the only one in the world who, however incapable of taking public exception to the exclusiveness of the title, might with reason deem that I had been unjust towards himself, in giving too great prominence in my book to the fact that it was the King's Rifle only which had saved the expedition, when he possessed an equal right to my gratitude, in having saved me in turn.

The original title, therefore, weighed upon my mind as an injustice, although it had been dictated solely by a contrary sentiment, being but little accustomed to burn incense on the altar of the great, and I immediately resolved to retain the title for the first part of my narrative, and give to the second part the name of François Coillard, the man who saved me, and, in doing so, saved the labours of the expedition which I directed. It was a simple act of duty on my part.

But this decision necessitated a general title for the work as a whole, no difficult matter to supply when a Continent has been crossed from sea to sea.

This is why my work is now called "How I crossed Africa."

I am sure that the title of a book of this kind can be but of trifling moment to the public. It is necessary to call it something, and I have given it the name under which it will appear.

I shall be exceedingly sorry if any one objects to it, but it cannot now be helped; it is fortunately not a matter of a nature to interfere with any man's slumber, and I trust it will not disturb or abridge mine.

LONDON, 61 GOWER STREET,
    12th December, 1880, at midnight.

CONTENTS TO VOL. I.

Part I.—THE KING'S RIFLE.

PROLOGUE.

	PAGE
I.—How I became an Explorer.	1
II.—Preparations for the Expedition. . .	12

CHAPTER I.

IN SEARCH OF CARRIERS.

Arrival at Loanda—The Governor Albuquerque—No carriers—I proceed to Zaire—Ambriz—I reach Porto da Lenha—Ransomed slaves—I hear of Stanley's arrival—I go to Kabenda—I take Stanley on board the *Tamega*—The officers of the gunboat—Stanley my guest—Our itinerary—Arrival of Ivens 17

CHAPTER II.

STILL IN SEARCH OF CARRIERS.

The Governor, Alfredo Pereira de Mello—The Governor's house—Things for which the government of the mother country is not responsible—A sketch of Benguella—Its trade—I am robbed—Another robbery—The Katembela—I obtain carriers—Arrival of Capello and Ivens—Fresh alteration of route—Another difficulty—Silva Porto, the old country trader—New obstacles—Capello goes to the Dumbo—Departure—The Dumbo—Fresh difficulties—Final start 31

CHAPTER III.

THE STORY OF A SHEEP.

Nine days in the desert—Want of water—The *ex-chefe* of Quillengues—I lose myself in the bush—Two shots in time—A little nigger and a negress missing—Loss of a donkey—Quillengues at last—Death of the sheep 52

CHAPTER IV.

THROUGH SUBJUGATED TERRITORY.

Journey to Ngóla—The native king Chimbarandongo—Beauty of the country—Arrival at Caconda—José d'Anchieta—No correspondence—Arrival of the *chefe*—We follow the carriers—Ivens goes to the Cunene, and I go to the Cunene—Return from Bandiera's house—Carriers wanting—My opinion 65

CHAPTER V.

TWENTY DAYS OF PROFOUND ANXIETY.

I leave Caconda—The native chief Quipembe—Quingolo and the chief Caimbo—Forty carriers—Fevers—The Huambo—The native chief Bilombo and his son Capôco—Eighty Carriers—Letters and news—All but lost!—I move onwards—A knotty question in the Chaca Quimbamba—The rivers Calaé, Canhungamua, and Cunene—A fresh and serious question in the Sambo country—The Cubango—Rains and storms—Serious illness—A terrible adventure—The Bihé at last! 83

CHAPTER VI.

BELMONTE.

In the Bihé—Severe Illness—Improvement—Belmonte—I determine to start for the Upper Zambesi—Letters to the Government—How the Expedition was organised in the Bihé—Difficulties, and how they were overcome—Historical and social notes on the Bihé—My labours—New difficulties—I leave Belmonte—The road to the Cuanza—Slavery 146

RAPID RETROSPECTIVE GLANCE . . 216

CHAPTER VII.

AMONG THE GANGUELLAS.

Passage of the Cuanza—The Quimbandes—The Sova Mavanda—The rivers Varea and Onda—Tree-ferns—Tribulations—Slaves—The river Cuito—The Luchazes—Emigration of Quibocos—Cambuta——The Cuando—Leopards—The Ambuellas—The Sova Moem-Cahenda—Descent of the river Cubangui—The Quichobos—Sudden changes—I start for the Cuchibi 226

CHAPTER VIII.

THE KING OF THE AMBUELLA'S DAUGHTERS.

The Cuchibi—The Sova Cahu-heú-úe—The Mucassequeres—Opudo and Capéu—Abundance—Kindness of the Aborigines—Peoples and Customs—A Ford of the Cuchibi—The river Chicului—Game—Wild Animals—The river Chalongo—An awful day—The Sources of the Ninda—The Tomb of Luiz Albino—The Plain of the Nhengo—Labour and Hunger—The Zambesi at last! 301

LIST OF ILLUSTRATIONS.

(VOL. I.)

FIG.		PAGE
1.	Mundombe Women, Vendors of Coal. (*From a photograph by the chemist Monteiro*.)	47
2.	Mundombe Women and Girls (*From a photograph by Monteiro*.)	49
3.	Mundombe Men (*From a photograph by Monteiro*.)	50
4.	Man and Woman of the Huambo	96
5.	Woman of the Sambo	110
6.	My Encampment between the Sambo and the Bihé	111
7.	Cassanha Bridge over the River Cubango	116
8.	The Seculo who gave me a Pig	117
9.	Ganguella Women on the Banks of the Cubango	120
10.	Ant-hills on the Banks of the River Cutato of the Ganguellas	to face 122
11.	Ant-hill 13 feet high on the Banks of the River Cutato dos Ganguellas, covered with Vegetation	122
12.	Tomb of a Native Chief	124
13.	Caquingue Blacksmiths	127
14.	1. Bellows; 2. Clay Muzzle; 3. Anvil; 4. Hammer	128
15.	Articles manufactured by the Natives between the Coast and the Bihé	129
16.	Belmont House, Bihé	to face 149
17.	View of the Exterior of the Village of Belmonte in the Bihé	150
18.	Plan of the Village of the Belmonte in the Bihé	151
19.	Woman of the Bihé, digging	161
20.	Biheno Carrier on the March	162
21.	Simple Palisade; Palisade bound together with withes; Palisade with forked uprights	176
22.	Plan of a native Libata or fortified village in the Bihé. Trophy of the chase found in almost all fortified villages	177
23.	Post erected outside the gate of the villages	177
24.	Articles manufactured by the Bihenos	185
25.	Quinda, or straw basket which will hold water; Large Sieve for drying rice or maize flour: Sifting Sieve; Ladle for watering the Capata	187

LIST OF ILLUSTRATIONS.

FIG.		PAGE
26.—A Bihé Head-Dress		190
27.—Bihé Women pounding Maize		200
28.—Ganguella, Luimba and Loena Women—Method of shaping the Incisors.		208
29.—Ant-Hills, found between the Coast and the Bihé		222
30.—Crossing the Cuanza		226
31.—Quimbande Man and Woman		227
32.—Quimbande Girls		228
33.—The Bihenos constructing Huts in the Encampments		232
34.—Skeleton of a Hut		233
35.—Hut built in an hour		234
36.—The Sova Mavanda, masked, and Dancing in my Camp	to face	238
37.—Quimbande Woman carrying her load		238
38.—1. Pipe; 2. Knives; 3. Tomahawks		240
39.—Ditassoa—Fish of the River Onda		245
40.—Tree-ferns on the Banks of the Onda		246
41.—Cabango Woman's Head-Dress		247
42.—Cabango Man		248
43.—Cabango Man		249
44.—Lake Liguri		253
45.—A Luchaze of the Banks of the River Cuito		255
46.—Tinder-box, Flint and Steel		256
47.—A Luchaze Woman on the Road		257
47A.—Atundo, Plant and Fruit		269
48.—Village of Cambuta, Luchaze		274
49.—Luchaze Woman of Cambuta		275
50.—Luchaze Man of Cabuta		276
51.—Articles manufactured by the Luchazes		277
52.—Luchaze Woman of Cutangjo		280
53.—Luchaze Pipe		281
54.—Luchaze Fowl-house		281
55.—The *Urivi*, or Trap for small Game		282
56.—Luchaze of the Cutangjo		283
57.—Luchaze Articles		283
58.—The Cuchibi		287
59.—Leaf and Fruit of the Cuchibi		288
60.—The Mapole, Tree and Leaf		289
61.—Mapole, Fruit and arrangement of the Branches		291
62.—Moene-Cahenga, Sova of Cangamba; 1. Fly-flap		293
63.—(Chimbenzengue.) Hatchet of the Ambuellas of Cangamba		294
64.—Ambuella Pipe		295
65.—The Quichôbo	to face	299
66.—The Oúco		305
67.—The Opumbulume		306
68.—Rat		308
69.—The Songue; Slot of the Songue		311
70.—The Sova Cahu-heú-úe		318

LIST OF ILLUSTRATIONS.

FIG.		PAGE
71.—Ambuella Woman		325
72.—Opudo		326
73.—Capéu		328
74.—Cuchibi Canoe and Paddle		331
75.—Drum used at Ambuella Feasts		332
76.—Caú-eu-hue (Town on the Cuchibi)	to face	333
77.—The Sova's Brother		335
78.—Ambuella Hunter		340
79.—Chinguêne		341
80.—Lincumba		342
81.—Chipulo or Nhele		342
82.—The Cuchibi Ford	to face	343
83.—Assegais of the Ambuellas		345
84.—Ambuella Arrow-Heads		346
85.—Malanca		353
86.—1. Direction of horns seen from the front; 2. Slot of the Malanca		354
86A. The Buffalo	to face	360
87.—Luina Shield		369
88.—The Chief Cicóta		370
89.—Ant-hills of the Nhengo		371
90.—1. and 2. Luina Houses; 3. Granary; 4. Luina Hoe		372
91.—Vertical Section of a Luina House in the Village of Tapa		374

MAPS IN VOL. I.

Map No. 1.—Tropical South Africa		In pocket
,, ,, 2.—Benguella to the Bihé	to face	216
,, ,, 3.—Cubango to Cuanza	,,	218
,, ,, 4.—Country of the Quimbandes	,,	230
,, ,, 5.—Disposition of the water at Cangala	,,	273
,, ,, 6.—Cambuta to the River Cubangui	,,	279
,, ,, 7.—Marsh of the source of the Cuando	,,	285
,, ,, 8.—From Cangamba to the Cuchibi	,,	316

HOW I CROSSED AFRICA.

Part I.—THE KING'S RIFLE.

PROLOGUE.

I.—How I became an Explorer.

In the course of the year 1869 I formed part of the column which, in the Lower Zambesi, sustained many a sanguinary conflict with the natives of Massangano. Senhor José Maria Latino Coelho, the then Minister of Marine, gave orders to the Governor of Mozambique, to furnish me, at the close of the war, with the means of mounting the Zambesi, so that I might make a detailed survey of as much of the country as it was possible for me to investigate.

The orders were given, but were never carried out; and after repeated applications and a hasty run through the Portuguese possessions of Eastern Africa, I returned to Europe, with a greater desire than ever to study the interior of that continent of which I had obtained only a superficial glance.

Private reasons of a family nature stepped in to defer and, even for a time, to destroy my projects.—

An officer in the army, always in garrison in small provincial towns, I was accustomed to convert my hours of idleness into hours of labour; and though it appeared to me that the possibility of visiting Africa was remote, the study of African questions became my sole and exclusive pastime.

Nor did I neglect the sublime subject of Astronomy, so that the abundant leisure which my barrack life allowed me was equally divided between Africa and a study of the heavenly bodies.

In 1875 I was in the 12th Chasseurs, and had in my comrade, Captain Daniel Simões Soares, one of the most intelligent men it has ever been my fate to know. We had not been acquainted long ere we became fast friends.

The wretched little room of this illustrious officer in the barracks of the Island of Madeira gave us mutual shelter during the hours that the regulations compelled us to reside there; and how often, when one of us was on guard, did he not have the other for companion! Africa, and still Africa, was our subject of conversation. It is a pleasure to recall that time, those hours which fled so rapidly, discussing questions which I was far from thinking I should be one day called upon to solve.

Towards the close of 1875 I drew up a paper which I submitted to the judgment of Simões Soares and of another of my comrades, Captain Camacho, and which owed its origin to our interminable talks about Africa. In it I laid down a plan for a partial survey of the interior of our colonies in East Africa, which might be effected with the greatest economy to the State.

After the question had been discussed and rediscussed among us, the paper was forwarded to His Majesty's Government: but I learned subsequently that it never reached the hands of the Minister of Marine.

At that time I was again revolving in my mind a return to Africa, notwithstanding my ties as a family man and the great personal interests which attached me to Portugal.

About the end of 1876 I returned to Lisbon, where I learned that African matters had assumed considerable importance in that city owing to the creation of the Central Permanent Geographical Commission and the establishment of the Geographical Society of Lisbon.

There was especially much talk about a great geographical expedition to the interior of Southern Africa.

I at once set about seeing the Minister of the Colonies, Snr. João d'Andrade Corvo. If it be no easy matter to explore Africa, it is scarcely less difficult to get an interview with a minister, more especially if that minister be like Snr. João d'Andrade Corvo. His Excellency held two portfolios, Marine and Foreign Affairs, and it may be conceived that he had no time to bestow upon intruders. I hunted him up for eight days in succession, and on the very eve of my departure from Lisbon I obtained an audience at the Ministry for Foreign Affairs.

His Excellency received me somewhat stiffly, observing that he had but little time to dispose of; he then inquired what I wanted of him.

This question led to the following dialogue:—

"I have heard it stated that Y. E. is thinking of sending a geographical expedition into Africa; and that is the object of my calling."

The minister immediately changed his tone, and very courteously desired me to be seated.

"Have you been in Africa?" he asked.

"I have; I know something of the mode of travelling in the country, and have devoted much attention to the study of African questions."

"Do you feel inclined to make a long journey into South Central Africa?"

I must declare that I hesitated a moment before replying; but at length I said—

"I am ready to go."

"That is well," he observed. Then he continued: "I have thoughts of sending out a great expedition to Africa, well provided with all necessaries, and when the organisation of the staff is under consideration I will not forget your name."

"By-the-by," he said, when I was on the point of leaving, "what terms do you ask for such a service?"

"None," I replied. And so we parted.

From the Ministry for Foreign Affairs I went to No. 3 Calçada da Gloria, to call upon Dr. Bernardino Antonio Gomes, Vice-President of the Central Permanent Geographical Commission. We had a long conference together, and that distinguished scholar, who was then entirely devoted to geographical subjects, told me that he had already cast his eye upon a distinguished officer of our royal navy, Hermenigildo Capello, to form a part of the expedition.

On the following day I started for the north. The journey and the fresh air of the country somewhat chilled the feverish enthusiasm which had taken possession of me in Lisbon, and, after mature reflection, I resolved to give up exploring Africa.

My wife and daughter were difficult ties to rend asunder, and whenever the idea of tearing myself from the tender caresses of my child crossed my mind, the ardour of exploration gradually died within me.

My family on the one hand and Africa on the other pulled my heart-strings in opposite directions, and kept me a long time in a state of perplexity. I at length hit upon a scheme which I thought might solve the question. Were I, for instance, appointed to the governorship of a district, I might make a portion of Africa my study without separating myself from my family. I was then serving in the 4th Chasseurs, and on my journey to Algarve I spent a few days in Lisbon. An exploring expedition was no longer talked of, and but one enthusiast, Luciano Cordeiro, still held to the belief that it would be brought about; although in the Geographical Society, of which I was the Secretary, a loud cry had been raised in its favour. Dr. Bernardino Antonio Gomes, bowed down by the weight of years, had yielded to the pressure of his incessant labours, and already felt the first symptoms of that disease which, a little later on, deprived him of his valuable life, and snatched from Portugal, and the world at large, one of the most illustrious Portuguese of the nineteenth century.

I was not at that time acquainted with the ardent and brilliant youth for whom I feel to-day so warm a friendship—I mean Luciano Cordeiro.

All those with whom I conversed of exploration told me it must be looked upon as adjourned *sine die*. Although the state in which I found matters at Lisbon caused me poignant regret, seeing that the light which had at one moment burned to give so harmonious an impulse to Portuguese exploration in Africa appeared to be flickering—on the other hand, I could not but feel a certain pleasure at finding myself, by this course of events, freed from an engagement which would have separated me from beings I held so dear.

The idea of going out as a governor and of establishing myself in Africa,—in that continent where I so ardently desired to labour and yet not separate myself from my family—became stronger every day, and I at length waited upon the minister to broach the subject.

This time I was received at once, and very cordially too. I expressed my surprise at hearing no more about explorations.

" And that has brought you here?" was the inquiry.

"Not exactly. I have come to entreat of Y. E. the governorship of Quillimane, which is now vacant."

Snr. Corvo smiled. "I have a mission of far higher

moment to entrust to you," he said; "I want you for a very different matter than to govern an African district; so that I cannot give you the governorship of Quillimane."

"Y. E., then, is still thinking of an African exploration?" I replied. "Frankly, I believed that the whole thing was at an end."

"I give you my word of honour," said the minister, "that either I shall cease to be João d'Andrade Corvo, or next spring an expedition, organised in a way hitherto unknown in Europe, shall leave Lisbon for South Central Africa."

"And you count upon me?"

"I do most certainly—and you will very shortly hear from me."

I left the ministerial presence in a state of bewilderment.

On arriving at the Hôtel Central, I sat down and wrote the following note:—

"I have not the honour of your acquaintance, but I wish to speak with you, and beg that you will favour me with an interview."

This I addressed to "Hermenigildo Carlos de Brito Capello—Officer on board the plated frigate *Vasco da Gama*."

The very next day I received the following reply:—

"You will find me to-day, at 3, at the Café Martinho. —Capello."

As the clock struck three I entered the Café Martinho, to find the place completely empty. No, not completely, for at one of the tables sat a young man in the uniform of a first lieutenant in the navy, whose face was completely unknown to me. This, however, I thought, must be my man. He was leisurely sipping a glass of grog, his cap lying by his side.

He was of medium stature, so far as I could judge,

he being seated; had a swarthy complexion, and a singularly placid eye. The thinness of his hair, from which the colour had begun to fade, and a small moustache already tinged with grey, gave him, at a first glance, an appearance of age, which was belied by his look and the unwrinkled aspect of his skin.

"Snr. Capello, I presume?"

"That is my name; and you, I suppose, are Snr. Serpa Pinto? I was expecting you, and feel pretty sure that you wish to have some talk about Africa?"

"Exactly so. You have then decided to take part in the expedition?"

"I have; in fact, I have already had some conversation on the matter with Dr. Bernardino Antonio Gomes."

"It was he who mentioned to me your name. What engagement have you made?"

"None. To tell the truth, I do not well know what the Government want; I have spoken twice about the matter to Dr. Gomes, but have not yet seen the minister; when I do I wish to tell him that if I go to Africa I should like to have as a companion my friend and comrade Roberto Ivens. Do you know him?"

"I do not. I have spoken to the minister upon the subject, and he has told me that he counts upon me for the expedition."

"In that case, as you are under engagement to the minister, I shall cry off."

"But why so? . . . I would rather do so myself."

"Apart from this, I do not think the matter will ever be brought to bear."

"Nor do I entirely; but admitting that it is carried out, why should we not both go? We are new acquaintances, it is true; but more intimate relations will follow, and, as I believe, may end in close friendship."

"I myself see no reason to the contrary. If the expedition goes forward then, we will start together, and get my friend Roberto Ivens to join us."

"By all means. But do you seriously think the Government will vote so large a subsidy as will be necessary for such an undertaking as is contemplated?"

"I do not know; I doubt it; and just now the expedition is far less talked about than it was."

Our conversation lasted long, and when we separated it was with the firm conviction that the venture would never be realised.

I met Capello several times during the succeeding days, and when we finally parted it was to pursue our respective duties; he to join his ship the *Vasco da Gama*, ordered to England, and I to take the command in the Algarve of my company, the 4th Chasseurs.

With the leisure afforded me by my garrison life, I once more resumed my studies, and was fortunate enough to find at my new station a good friend in the person of Marrecas Ferreira, a distinguished engineer officer who was ever ready to assist me in difficult mathematical problems, of which he was a master. It was through him that I was enabled to enter into a regular correspondence with Luciano Cordeiro, who afterwards became one of my firmest friends.

It was during this time that I drew up two small papers, wherein I discussed the mode of organising an exploring expedition into South Central Africa, and, thanks to Luciano Cordeiro, they found their way into the hands of the Minister of Marine.

Still months passed away, and no more was heard of the promised expedition.

I received two letters from Capello, wherein he expressed his complete want of faith in the realisation of the undertaking. It is true that in the Permanent Geographical Commission various projects of expeditions were discussed, but they led to no action, and the matter appeared to be dying out.

One morning I read in the newspapers that the minister Snr. João d'Andrade Corvo had brought before Parliament a Bill for a credit of 30 contos (some £6600) for an expedition to Africa; but shortly after, before the Bill had passed, the ministry was defeated, and the Portfolio of the Colonies fell to Snr. José de Mello Gouvea.

The projected exploration, however, again became a subject of public interest; but the newspapers mentioned as explorers men who were totally unknown to me, and only occasionally mentioned the name of Capello.

I was then residing at Faro, and although I had not given up my astronomical and African studies, which I pursued with the assistance of João Botto, an eminent professor of the school of Pilots of Faro, I had ceased to cherish my former ideas of travel. My time was divided between home pleasures and my books of study, and I found myself too happy in the comforts of the domestic hearth to think of exchanging the even tenor of my life for the shocks and chances of a journey through savage climes.

Nevertheless, in my quiet retreat, I followed with interest the reports published in the journals of the news from Lisbon. I there read that the new minister, José de Mello Gouvea, had again brought before Parliament the Bill that had been introduced by his predecessor, João d'Andrade Corvo, and had succeeded in obtaining a vote for the sum of 30 contos, to be expended in an exploring expedition.

The death of Bernardino Antonio Gomes, a victim to the deep interest he took in the study of African questions, at an age when the fatigues of many previous years should have counselled him complete rest of mind, occurred about this time, and produced a great void in the Central Geographical Commission; so great indeed that although there were many of its members who, deeply interested in the subject, raised their voice in that learned body, their discussions led to no practical result.

In spite of the parliamentary vote, I could not satisfy myself that there was any possibility of seeing the expedition carried into effect in 1877; and bearing in mind what I had read in the public prints, I deemed, at least, that I was myself forgotten—a circumstance which, to tell the truth, was rather agreeable to me than otherwise.

The Algarve is a delicious country; a perfectly eastern atmosphere pervades the place, and seeing the elegant tops of the palm-trees gracefully bending over the terraced houses, one felt inclined at times to forget that one was still bound to the prosaic shores of Europe. My position there was that of military commandant, by which will be understood that my life was not a particularly hard one.

The intercourse of a select society, family affections, my books of study and scientific instruments, enabled me to spend very happy hours—of that placid happiness which it is not the fate of many to enjoy. My easy-chair, my dressing-gown and slippers, were fast becoming my very ideal of felicity.

April had come to an end, and with the beginning of May set in the heat, which was very powerfully felt in Faro. I began to form projects for the summer, when one day I received a telegram requiring me to report myself immediately to the general in command of the division. On proceeding thither I found an order to repair without loss of time to the presence of the Minister of the Colonies.

Adieu to home, adieu to dressing-gowns, adieu to

slippers! Adieu to the tranquil and placid life I had been spending amidst my dear ones! I must return to the busy world once more!

Four days later on, seated around a large table, in a great hall at the Ministry of Marine, were a dozen grave personages, some with spectacles and some without, some old and others new, but all well known in the scientific or literary world, or for their public services, who had met together to discuss the often mooted question of Africa. This solemn session was presided over by the minister José de Mello Gouvea.

The secretaries were, Dr. José Julio Rodrigues and Luciano Cordeiro, and I remember that among those present were Count de Ficalho, Marquis de Souza, Dr. Bocage, Carlos Testa, Jorge Figaniere, Francisco Costa, Counsellor Silva, and Antonio Teixeira de Vasconcellos.

At the bottom of the table, and at one of the corners, ensconced in a large fauteuil, was a man with a head well covered with hair, and a heavy grey moustache, who, through his tortoise-shell-rimmed glasses, kept his eyes steadily fixed upon me. It was the late minister, João d'Andrade Corvo, whose look said as plainly as words could do, "I told you that this matter would be brought to bear."

Capello sat next to me, and after a debate of some two hours we left the hall together, with precise instructions for our journey. We selected as third associate Lieutenant Roberto Ivens, Capello's friend above alluded to, who was unknown to me, and who was at that time at Loanda, serving on board one of His Majesty's vessels. It was on the 25th of May that the meeting was held, and we undertook to start on the 5th of July. It was a risky thing to promise, as we had to fit out the expedition in France and England, and we only had a month to do it in.

We received powerful assistance, however, at the hands of Francisco Costa, Director General at the Ministry, who used his influence to sweep away all the obstacles which the red-tapeism of the offices might conjure up, and in fact matters were so managed that on the 28th of May Capello and myself were enabled to leave for Paris and London, in order to make the necessary purchases. To this end, we were armed with a credit of eight contos, or about £1760 sterling.

II.—Preparations for the Expedition.

On our arrival in Paris, we called upon M. d'Abbadie, the great Abyssinian explorer, and also upon M. Ferdinand de Lesseps, from both of whom we obtained advice and received the politest attention.

Unluckily we could not find in the market any of the instruments, arms, or travelling appointments such as we desired, so that it became necessary to order the whole of them.

Backed by a special recommendation from M. d'Abbadie, we applied to various instrument-makers, and during some ten or twelve days Lorieux, Baudin, and Radiguet were hard at work for us. Walker undertook to supply all the travelling requisites; Lepage (Fauré), the arms; Tissier, the boots and shoes; and Ducet jeune, the body clothing.

Our Paris orders being thus well in hand, we started for London, where we purchased our chronometers, of the firm of Dent, and sundry instruments of Casella; a good store of sulphate of quinine was also laid in, and many india-rubber articles were procured from Mackintosh, among others being two boats and some folding baths.

We sought in vain in London, just as we had sought

with like ill-success in Paris, for a theodolite possessing the necessary conditions for a journey such as that we were about to undertake. Some, that were excellent for terrestrial observations, were wanting in those requisites which astronomical observations demanded; others again, that were perfectly satisfactory in both respects, were either too heavy or too bulky for our purpose. There was no time to have a special one made for us, so that on our return to Paris we were glad to accept one that had been previously offered us by M. d'Abbadie.

We collected together, in Paris, the various articles we had ordered and that had been made during our short absence from that city; and on the 1st of July, Capello and myself arrived at Lisbon completely prepared for our journey, and consequently ready to fulfil our engagement to leave for Loanda by the packet of the 5th. Our preparations had been made in the space of nineteen days.

When I was studying the means of preparing myself for a long journey in Africa, I procured various books of travels, in the vain hope of gleaning from their pages the modes of preparation adopted by other travellers. All the narratives were, however, singularly wanting in information of this kind, and remembering the degree of annoyance which the omission caused me, I resolved that if I should ever travel in Africa and write an account of my adventures, I would supply the deficiency, and, whilst enumerating the articles I took with me, I would put on record which among them proved of real service, and which might be considered as mere lumber.

The story of African exploration is in its early infancy. Many explorers will succeed me in Africa, as I succeeded others, and I believe that I shall be doing a service to those who venture after me on to the inhospitable continent, by furnishing them with a list of the things with which I provided myself, and they will glean from the course of my narrative the advantages or inconveniences I found from their employment.

According to the instructions I received from the Government, I was at liberty to expend three years upon my journey, and it was upon this understanding that I made my preparations.

Experience had taught me the serious inconvenience of overloading myself with baggage, and I therefore frankly declare that when I surveyed in Lisbon the enormous pile of things purchased in Paris and London, I was perfectly horrified. There were no fewer than seventeen trunks! all of the same dimensions, 0m., 3 × 0m., 3 × 0m. 6.

One contained a toilet service complete, with a large mirror, basin and ewer, soap and brush dishes, etc.; another held a dinner and tea service for three persons, and a third the necessary kitchen utensils.

Three other trunks of extra strength were destined each to contain the following: four bottles of quinine, a small medicine-chest, a sextant, an artificial horizon, a chronometer, various tables of logarithms, some ephemerides, an aneroid, a hypsometer, a thermometer, a prismatic compass, a plain compass, a book of blank paper, pencils, loose paper and ink; fifty cartridges for each firearm, a complete suit of clothes, and three changes of linen, tinder, flint and steel, and other small articles for personal use.

Each of these trunks had a tray in the upper part containing a writing-case and place for paper. They were considered as personal luggage, and belonged each to one of our party.

The remaining ten trunks were packed indiscriminately with articles of clothing, instruments, and other matters in reserve. The locks of all of them were the same, and one key opened the whole.

Our tent was of the kind known as a *tente marquise*, 9 ft. 4 in. wide by 6 ft. 3 in. in height. The bedsteads were of iron, strong and convenient; the tables were folding, the stools and chairs of canvas. All these articles were manufactured by Walker.

Each of us was armed with a magnificent rifle of sixteen-bore, the barrels of which—the work of Leopold Bernard—had been carefully mounted by Fauré Lepage. A fowling piece of the same calibre, manufactured by Devisme, a Winchester eight-shooter, a revolver and a wood-knife, completed our armament.

I had ordered at Lisbon, of the Confeitaria Ultramarina, twenty-four cases, of the same dimensions as the trunks, to be packed (in tins, carefully soldered) with tea, coffee, sugar, dried vegetables, and farinaceous substances; and I must here express my warmest thanks to Snr. Oliveira, the proprietor of that establishment, for the scrupulous care he bestowed on the selection of the articles supplied, and of which we made great use at the outset of our journey.

The instruments we carried with us were the following: three sextants, one made by Casella of London, one by Secretan, and the third by Lorieux, a perfect beauty; two Pistor's circles, manufactured by Lorieux, with two glass horizons and the respective levels; one Secretan's mercury horizon; three astronomical telescopes of great power, two by Bardou and one by Casella; three small aneroids, two of Secretan's and one of Casella's; four pedometers, two of Secretan's and two of Casella's; six algebraic compasses; one Bournier's compass, furnished by Secretan; three others, azimuths, two from Berlin and one supplied by Casella; two of Duchemin's circular needles; six Baudin's hypsometers, one of Casella's, three of

Celsius of Berlin, two others of Baudin's highly sensitive; twelve thermometers, supplied by Baudin, Celsius, and Casella; one Marioti-Casella's barometer; one Casella's anemometer; two Bardou's binoculars; one dipping needle; and an apparatus of magnetic force, most kindly lent us by Captain Evans through the instrumentality of M. d'Abbadie. And finally, d'Abbadie's universal theodolite, to which has been given the name of *Aba*, and which its inventor so generously placed at our disposal.

Arms, instruments, baggage, in a word, all the articles we took with us, bore the following inscription—*Portuguese Expedition to Interior of Southern Africa*—1877.

Two cases containing the needful for the preservation of zoologic and botanic specimens were forwarded to us by Dr. Bocage and Count de Ficalho.

Tools of various kinds swelled this enormous mass of impedimenta with which we were about to leave Lisbon in order to plunge into the unknown regions of South Central Africa.

CHAPTER I.

IN SEARCH OF CARRIERS.

Arrival at Loanda—The Governor Albuquerque—No carriers—I proceed to Zaire—Ambriz—I reach Porto da Lenha—Ransomed slaves—I hear of Stanley's arrival—I go to Kabenda—I take Stanley on board the *Tamega*—The officers of the gunboat—Stanley my guest—Our itinerary—Arrival of Ivens.

On the 6th of August 1877 we arrived at Loanda, on board the steamer *Zaire*, under the command of Pedro d'Almeida Tito, to whom I am happy to offer here a tribute of gratitude for the favours he bestowed upon me during the voyage.

From the moment of quitting Lisbon, there was one thing that constantly occupied and worried my mind. Our luggage was enormous, and had still to be greatly increased in the shape of merchandise, beads, and other articles that were to be our money in the interior of the country.

In all works of travel dealing with this part of the African continent, I had read of the difficulties which many explorers had met with through the impossibility of obtaining a sufficient number of carriers for the indispensable transport of their baggage. How was I to obtain them? I learned at Cape Verde that a letter addressed by myself and Capello to Ivens never reached his hands, inasmuch as I there found by a telegram that Ivens was still at Lisbon, and could not possibly have attended to the injunctions we imposed upon him, to make a study of the question and see whether he could get us the necessary help at Loanda. An attempt also made at Cape Palmas turned out fruitless, and, notwithstanding the assistance rendered us by Captain Tito, not a single *Keruboy* could we obtain there.

We at length reached Loanda, and found hospitality under the roof of Snr. José Maria do Prado, one of the chief landowners and capitalists of the Province of Angola, who immediately placed at our disposal one of the many houses he possessed in the town, with accommodation sufficient to shelter the enormous equipage of the expedition.

We received much kind attention from Snr. Prado; and on the evening of the 6th we were waited on by one of the aides-de-camp of His Excellency Snr. Albuquerque, the Governor General, who sent us many cordial messages.

The next day, the 7th, we called upon his Excellency, and received a most friendly reception. The Governor was good enough to excuse the very undress attire in which I waited on him, for although the things I wore were capitally fitted for bush-life, they could scarcely be considered proper for a visit of ceremony.

Snr. Albuquerque, after assuring us that he would render us every assistance in the district under his government, concluded by pointing out the impossibility of obtaining for us the means of transport.

I fancy that there are few things more disagreeable to a traveller anxious to commence a journey into Africa, and with 400 loads of luggage to take with him, than to be told that *carriers are not obtainable*.

I at once determined to proceed to the northern part of the province, to see whether I could be more successful, and therefore begged Snr. Albuquerque to procure me a passage to the Zaire. The only war-vessel that could be placed at my disposal was then cruising in the mouth of that river; I resolved to join her, and to that end, on the 8th I started in one of the country boats, manned by eight blacks, supplied me by the captain of the port. I carried orders from the governor to the commandant of the gunboat. A voyage of 120 miles, in a small boat, with scarce room to stretch your legs, is anything but pleasant. To make matters worse, from Loanda to Ambriz I had nothing to eat but biscuits and sardines, for having resolved to start directly my boat was ready, I had no time to lay in any stores.

On the 9th, at daybreak, I arrived at Ambriz, a charming town, seated on the level summit of an eminence, with precipitous sides, that are washed by the sea, some 80 feet below.

The chief official was an employé of the Treasury, a Snr. Tavares, who showed me most marked attentions, as did indeed all the inhabitants of the town, and more especially Snr. Cordeiro, in whose house I was lodged.

At Ambriz I fell in with Avelino Fernandes, whose acquaintance I had been fortunate enough to make on board the steamer *Zaire*, and intimate relations sprang up between us. He was born on the banks of the Zaire, and has a perfect passion for that rich soil, whose gigantic trees, the offspring of a virgin forest, shaded his cradle. His age is 24. His swarthy complexion and crisp curly hair indicate that there flows through his veins some African as well as European blood. Wealthy, possessed of a cultivated mind, his education having been obtained in the chief capitals of Europe, and endowed with superior intelligence, he is a true type of the courteous gentleman, whom to know is to sympathise with and esteem. The numerous connections he possessed in Zaire might, I thought, assist me in arranging the difficult question of transport.

I learned at Ambriz that the gunboat *Tamega* was expected there in the course of a couple of days, and I therefore resolved to wait for her. My voyage from Loanda in the country-boat had not left such a pleasing impression on my mind or body as to induce me to continue with her further northwards.

On the 10th I took a ramble about the town and suburbs, and the following is a brief record of my impressions.

From the plateau on which the European population have installed themselves you descend to the sea-shore by a zigzag road, which was then being reconstructed by convict labour. On the beach, between two fine blocks of building, used as warehouses by French and Dutch commercial firms, there exists a large structure, partly in ruin through age, and partially in course of re-erection, but with its works abandoned. This is the custom-house—but a custom-house without merchandise, where the goods heaped up at the door, upon the sand, pay an absurd tribute for warehousage. N.N.E. of the town, many acres of land are taken up with a marsh, which is at least 10 feet deep when at the highest, and on the sides of the slope which leads from the plateau to this marsh are scattered the huts of the native population, under the very worst conditions from a sanitary point of view. South of the town, among clumps of virgin wood, is situated the cemetery, where the bodies interred during the day become the food of hyenas at night.

The landing wharf or pier, built of iron and timber, is falling into utter uselessness; for, as it has never seen a coat of paint, as there is no fund to keep it up, and it is nobody's business to look after it, the natural result is that the iron, in rusting through contact with the air and water, is fast consuming both itself and the woodwork together.

The residence of the chief official is little better than a barn, which it is truly dangerous to dwell in. The powder-magazine is in no better condition, a fact which somewhat surprised me, as it contains the powder of the trade, producing no less a revenue to the State than two hundred milreis per month.

It is greatly to be hoped that during the two years which have elapsed since my visit to Ambriz, a little more care has been bestowed upon that pretty town, whose importance as a great commercial centre is patent to the least observant eye.

At the distance of about three-fifths of a mile N. of the landing-place, the river Loge debouches into the Atlantic. Its mouth is obstructed by a sandbank, which renders the river difficult of access, but, this passed, it is navigable for nearly twenty miles.

On the 11th I paid a visit to the important agricultural establishment founded by the celebrated Jacintho, known as Jacintho do Ambriz, and now the property of his son Nicoláo. This estate is one of the most remarkable in the province of Angola for the development of agriculture.

Jacintho do Ambriz found his way to Africa through a great calamity. A son of the people, without the slightest instruction, not knowing even how to read or write, but endowed with a clear understanding, a keen perception, and a happy temperament, he succeeded in realising a large fortune. Jacintho married in Ambriz a woman in his own class of life. She was the Tia Leonarda (*Aunt* Leonarda), better known as *Tia Lina*, a native of Beira-Alta; and in 1877 I remember her always dressed in the costume of the Beirense peasantry, talking the corrupt language spoken by the people of that province. I remember, too, being regaled in her house with a Beirense dinner, and for the moment I imagined myself transported back among our agriculturists of the north. Tia Lina, from her energy and thrift, had a great share in assisting Jacintho to his handsome fortune.

Jacintho was engaged in trade, and that trade in Africa was divisible into two branches, viz. the purchasing goods from the whites and selling them the produce of the country, and purchasing such produce from the blacks and selling to them the aforesaid goods. It was Jacintho's part to treat with the whites, and Tia Lina's with the blacks.

Jacintho, who was a generous-hearted fellow, too often fell a victim to his own honesty and the extortions of some of the chiefs—a fact which often drew from Tia Lina the expression, "Ah, Jacintho, the whites bamboozle you; but I bamboozle the blacks!"

The verb employed by Tia Lina was not precisely *bamboozle*, but it will serve to express her meaning.

One day Jacintho took it into his head to turn his attention to agriculture. It was the instinct of his early youthful habits working upon him. He purchased land, and laid the foundations of that vast estate which is fully worthy of a visit; to this, his hobby, he devoted his labour and care till the last moment of his life.

Jacintho was known for his strange misuse of words, and many curious stories are told of the droll mistakes he fell into through the wrong employment of this or that expression with which he larded his discourse, evidently unaware of the real meaning of the phrases he adopted. He had, however, a keen mother wit, and the laughter evoked was not always against him. Take for example the following anecdote.

He had already been settled for some time on his Logé estate, but on the arrival in the harbour of a Portuguese vessel of war, he went on board in the old style to offer things for sale to the officers. His genial nature made him always welcome, and he at once became familiar with officers and crew. One day the commander, seeing him on deck, asked him for a monkey. "How many do you want?" inquired Jacintho. "You may send a boat off to my house at Loge to-morrow morning, and fetch as many as you like."

He was taken at his word, and on the following day a boat, manned with half a dozen sailors, ran alongside Jacintho's garden-wall. The old fellow made the men row the boat a mile or so higher up until they reached the slope of a hill covered with gigantic baobab-trees, upon whose horizontal branches were swarming hundreds of monkeys. Turning to the sailors, Jacintho exclaimed, "There they are—they're all mine—catch as many as you like, and take 'em to the commandant with my compliments."

The men looked askance at the lofty tops of the enormous trees, whose trunks were too capacious for two or three of them to encircle, and after sundry vain efforts to scale the perpendicular height of these natural columns they gave up the task, amid the gibbering and chattering, which sounded very like derisive laughter, of the numerous monkey families.

"Don't say I didn't give 'em to you—there they are —you've only got to catch 'em," said Jacintho, accompanying each exclamation with a fresh burst of merriment, which seemed to awake an echo in the lofty branches above them.

I visited his estate, and could not but be struck with the fact that all his machines, apparatus, implements, etc., were of Portuguese manufacture. Jacintho would admit of nothing that was not Portuguese; and cost what it might, he procured all his articles, whether intended for agricultural or manufacturing purposes, from Lisbon.

The memory of this obscure man, better known for

the absurdities he uttered than the many excellent things he did, should be respected by all who are interested in the development of Africa; for he was a man who, in modern times, has done the very highest service in fostering agriculture in this Portuguese colony, where he employed his immense fortune, and where he personally laboured till the last day of his useful life.

On the left bank of the Loge is situated another agricultural property, also of importance, belonging to Snr. Augusto Garrido. I had not time to pay it a visit, as on the day I spent in that part of the district I could not escape the many kind attentions of Nicoláo and Tia Lina; and though I passed some hours there, they were all too few to examine and admire what the will of one man had been able to create out of the desert and the marsh.

The day following the one thus agreeably spent saw the arrival of the *Tamega* gunboat. I at once went on board, but found her without stores and with a large number of men on the sick-list; for which reason I arranged with the commandant, Snr. Marques da Silva, to wait for him at Ambriz whilst he went on to Loanda to recruit.

Three days later the *Tamega* came back, when I joined her, with Avelino Fernandes, and we immediately proceeded on our voyage to the Zaire.

I had been suffering for some time with acute bronchitis, which fortunately improved directly I found myself at sea.

We mounted the Zaire as far as Porto da Lenha, where I disembarked with Avelino Fernandes, who presented me to his friends in that place. I at once began to inquire about transport. They told me I might possibly obtain carriers if the native chiefs chose to assist me, but that the best plan would be to ransom a number of slaves and then engage them for the service I required. The idea of purchasing human flesh, although it might be with the view of setting them subsequently at liberty, was repugnant to me. And then, how could I tell whether they would stick to me after all, if once they were free?

I therefore determined to reject the notion, even if not a single carrier were to be had in the place.

I learned at the house where I was stopping that the great explorer Stanley had arrived at Boma on the 9th, having descended the entire course of the Zaire. He had come by the way of Kabenda.

I returned on board and arranged with the commandant to go on to Kabenda, to offer our services to the intrepid traveller. We set off at once, and were no sooner anchored in the roads than I went on shore with Avelino Fernandes and some of the officers of the gunboat.

I was quite affected as I pressed the hand of Stanley, who, though a man of small stature, assumed in my eyes the proportions of a giant.

I offered him my services in the name of the Portuguese Government, and told him that if he desired to go on to Loanda, where he could most easily obtain transport for Europe, Commandant Marques would willingly give him and his men a passage on board the gunboat. In the name of the Portuguese Government I further placed at his disposal the money he required.

Stanley answered me with a warm pressure of the hand.

The officers of the *Tamega* confirmed my offer in the name of their commandant.

Stanley accepted it, and from that moment the gunboat remained at his disposal.

As may well be conceived, neither I nor Avelino Fernandes allowed Stanley to go out of our sight, and, eager to hear the particulars of his journey, we utilised every moment in questioning him and his men.

On the 19th, the officers of the *Tameya* gave a splendid banquet to the great explorer, to which Commandant Marques invited Fernandes and myself.

On the 20th we set off for Loanda, having on board the whole of Stanley's followers, to the number of 114 persons, among whom were twelve women and a few children.

Stanley was lodged at Loanda in my own house—a distinction which was very agreeable to me, as he refused many other invitations, some from persons who could offer him accommodation far beyond my powers, seeing that the only furniture my poor dwelling contained was that supplied by my travelling resources.

The Governor immediately sent a messenger to compliment the illustrious American, and invited him to a banquet, at which I was present. On our way home, I asked Stanley what impression Snr. Albuquerque made upon him, to which he merely replied, "He seems a very cold kind of gentleman."

The American Consul, Mr. Newton, gave us a breakfast, and showed us much kind attention.

Other festivals and banquets followed; time was flowing on; we had reached the 23rd of August, and still not a single carrier had been obtained. It was in the evening of the dinner given by the Governor to Stanley that His Excellency repeated that it would not be possible for me to obtain transport at Loanda, and in support of his assertion referred to the case of Major Gorjão, who had scarce obtained half the men he wanted, when engaged on the survey of the Cuanza railway.

It is now time to speak of our projects, as defined by law, and the instructions of the Government.

Parliament, as has been stated, voted a sum of 30

contos of reis (£6600) for the purpose of surveying the hydrographic relations between the Congo and Zambesi basins, and the countries comprised between the Portuguese Colonies, on both coasts of South Central Africa.

Subsequent instructions laid more particular stress on a survey of the river Cuango in connection with the Zaire; a study of the countries in which the Coanza, Cunene, and Cobango take their rise, as far as the upper Zambesi; and, if possible, a careful survey of the course of the Cunene.

The plan as set forth in the Act of Parliament, which had been drawn up by Snr. Corvo, would appear at first sight far too vast a scheme for a single expedition and a vote of 30 contos of reis; but the Act nevertheless was carefully worded. Snr. Corvo was aware that not only is a traveller in Africa not always master of his actions, but is likely to meet upon his road with some unforeseen problem, the solution of which he may deem of far greater importance than the one he was sent over to study; and on that account great latitude was allowed the explorers.

As regards the instructions, they were more restricted, but even they by no means trammelled the movements of the expedition.

As to the point of entry, seeing that it depended essentially upon the most convenient spot for obtaining transport, it was left to our discretion.

Capello and myself had thought of making our entry at Loanda, travelling eastward until we reached the Cuango; descending that river for two degrees, entering the Cassbi, by which we intended to descend to the Zaire; and finally, investigating the Zaire to its mouth.

The arrival of Stanley, who had performed a part of the labour we had tracked out for ourselves, and above all the impossibility of obtaining carriers at Loanda, made us completely alter our plans.

We decided now that I should go southwards to procure some men in Benguella; and that, if I could obtain them there, we would enter by the mouth of the River Cunene, go up it to its source, and thence proceed in a south-east direction, as far as the Zambesi.

As no great confidence could be reposed in the men we hired, we thought it well to solicit the Governor for a certain number of soldiers, to act as a kind of escort. His Excellency acceded to the request, and passed the word among the regiments to learn whether any of the soldiers felt inclined to volunteer; for as the service was not a regular one, he could not compel any of the men to go.

It was therefore decided that I should start for Benguella by the steamer which would arrive from Lisbon about the beginning of September.

On board that steamer I met with our companion Ivens for the first time. Of a genial and ardent nature, possessing a great flow of words, and perfectly enthusiastic on the subject of difficult journeys, we soon became friends. We communicated to him all we had determined to do, and the difficulties we had hitherto met with. Ivens agreed with us as to the course to be adopted, and my departure for Benguella was definitely fixed for the 6th of the month.

I lost no time in getting ready for the voyage, and waited upon the Governor to apprise him of the circumstance. During my absence, my companions were to arrange and prepare the baggage, which, owing to our hasty flight from Europe, was in a state of considerable disorder.

I wish here to put on record an episode which annoyed me not a little, inasmuch as it might perhaps have led Stanley to form an estimate of my character and that of my companions far different to the true one.

On the 5th of the month, at breakfast, we were all

of us—Capello, Ivens, Stanley, and Avelino Fernandes
—talking about slavery, and we were explaining to
our guest the spirit of the Portuguese laws upon that
infamous traffic, seeking to impress upon him the
falsity of the assertions of foreigners in respect of our
country, and the impossibility of any slaves being held
where the government had any authority. In the
midst of the conversation Capello had to go off to the
Palace for an interview with the governor.

An hour later he returned, and very shortly afterwards Stanley received an official letter from Snr.
Albuquerque begging him to give a certificate to the
effect that " No slavery was permitted within the territory under his charge." Stanley, in a state of astonishment, showed me the letter, and most certainly both
myself and companions were as surprised as he. To
say the least of it, the thing looked very queer; and
our conversation at breakfast, followed up by the letter
so soon after the visit of one of us to the palace, might
appear to the illustrious traveller something outside
the pale of accident.

Stanley could undoubtedly certify to His Excellency
that neither on board the *Tamega* nor in any house,
neither at His Excellency's residence nor in that of
Consul Newton, had he seen any evidence of slavery.
But beyond this, as the Governor must have well
known, Stanley could have no information apart from
what he had obtained from us, and with the exception
of the town in which he was temporarily dwelling
he had visited no portion of the territory governed
by Snr. Albuquerque. To get from Stanley such a
document was to make him pay pretty dearly for a
dinner and other favours bestowed upon him. I
believe that Stanley did us the justice to think we
had no hand whatsoever in the conception or production of the letter.

On the 6th I left for Benguella, taking with me letters from Snr. José Maria do Prado to various private individuals, but without any recommendation to the governor of the district, with whom I was not acquainted.

I was once more about to search for carriers whom I, a Portuguese, had been unable to obtain in Loanda, and that four months later, a foreigner, the explorer Schutt, procured without difficulty, when pursuing the first route we had intended to follow.

On my voyage I made the acquaintance of a passenger who told me that I might possibly get a few carriers at Novo Redondo, and that he would undertake to contract there for some twenty or thirty of them.

This put me in rather better spirits, and it was in such humour that I arrived at Benguella on the evening of the 7th. Although I had letters of recommendation for various merchants, I determined to call upon the governor, and ask his hospitality, with what result the next chapter will show.

CHAPTER II.

STILL IN SEARCH OF CARRIERS.

The Governor, Alfredo Pereira de Mello—The Governor's house—Things for which the government of the mother country is not responsible—A sketch of Benguella—Its trade—I am robbed—Another robbery—The Katambela—I obtain carriers—Arrival of Capello and Ivens—Fresh alteration of route—Another difficulty—Silva Porto, the old country trader—New obstacles—Capello goes to the Dumbo—Departure—The Dumbo—Fresh difficulties—Final start.

ALFREDO PEREIRA DE MELLO,* Governor of Benguella, on hearing my request for hospitality, exhibited an amount of embarrassment which was only too perceptible, and after a pause said that he had no accommodation to offer me. His answer surprised me, as I knew him to be naturally courteous and open-handed. I had received invitations, from the very moment of my arrival, both from Antonio Ferreira Marques and Cauchoix, but I had made up my mind to take up my quarters in the Governor's house.

He then said that he had not a bed to offer me, at which I pointed to my travelling bed, for I had had my luggage brought up with me. Defeated in this quarter, he asserted that he had not a room; to which I responded by saying that a corner of the hall in which we stood would serve my turn.

Finding his objections thus overruled, he gave in,

* Alfredo Pereira de Mello, a captain in the army and Governor of Benguella, was the same Lieutenant Mello referred to by Cameron in his work 'Across Africa,' and who was then aide-de-camp to the Governor of the Province, Snr. Andrade.—*Note of the Author.*

and I stopped. I was curious to learn the cause of the Governor thus denying me hospitality, and a little investigation unravelled the mystery.

Alfredo Pereira de Mello was a new man, although he had attained to some rank in the navy. Congenial and intelligent, he was esteemed by all who knew him intimately, because to a finished education he joined a singular rectitude of character, and that energy which is peculiar to every good sailor. He had served in the English navy, and was an experienced navigator.

He had visited both the Americas, and before going to Africa in the capacity of aide-de-camp to the Governor Andrade, he had made voyages to India, China, and Japan.

His Excellency, who knew me very well by name, on hearing my request, forgot that he had the explorer before him, and only thought of the man, habituated to a life of comfort and even luxury. The truth therefore was, that Pereira de Mello was ashamed to offer me shelter.

A Governor of Benguella, however upright and honourable he may be, is bound to live in the very humblest fashion, if dependent on the pay that he receives.

The Government house is a hired one. Its furniture, many degrees below the designation of simple, is barely sufficient to garnish a sitting-room and one bed-chamber.

In the former, in striking contrast to the furniture, and in a richly gilded frame, was a portrait of the King, the best I have ever seen.

Foreign vessels of war frequently put into the harbour. The officers, on coming ashore, naturally called upon the Governor, and invited him on board, where they regaled him in right royal fashion, but not a glass of water did they get in return; and why?

because the negress who constituted the chief part of the domestic establishment of his Excellency would have had to present it on a cracked old plate. The so-called dinner-service was, I verily believe, like another sword of Damocles suspended over the head of Pereira de Mello, when I appeared before him and obstinately determined to remain his guest. And yet he was quite wrong. The neatness and cleanliness which presided at his board made you quite forget that the glasses were cracked and the plates chipped and otherwise disfigured by time, and the simple but admirably cooked food was so appetising after exposure to the air of Africa that—though I have no wish to offend the cook at the Hôtel Central in Lisbon—I must aver that I have dined better in the Governor's house at Benguella than ever I did off his savoury viands; and yet I will lay any odds that the negress Conceição, who performed such wonders of cookery, never even heard the name of that hero of pots and pans, the celebrated Brillat-Savarin.

The very first day of my forcible entry on his privacy, Pereira de Mello opened to me his heart and entered into many details of his inner life. Three official notes addressed to the government of the province, wherein he begged for authority to make certain reforms in his household, had remained, he said, unanswered.

How little novelty is there in human affairs! On turning over the leaves of a copy letter-book, existing in the archives of the government of Benguella, I happened to fall upon certain official notes dated as far back as 1790, wherein the then governor made an appeal to the king in almost identical terms; averring that he had complained in vain to the governor-general of the province about the state of the carriages of two brass guns, which urgently required looking to;—application and appeal having been both, alas! equally fruitless, as the pieces are carriageless at the present day!

These are the very pieces of ordnance alluded to by Cameron. He will be pleased, however, to know that the carriages have been ordered, and cannot be much longer delayed; for as the order appears to have been given at some time in the said year, 1790, they must surely be nearly ready now.

Benguella is a picturesque town which extends from the shore of the Atlantic to the very summit of the mountains which form the first steps of the lofty plateau of tropical Africa. It is surrounded by a dense forest, the *Matta do Cavaco*, even at the present day peopled with wild beasts; a fact, however, which should cause no particular astonishment, inasmuch as the Portuguese generally are not greatly given to sport. The residences of the Europeans cover a large area, for all the houses have vast gardens and dependencies. These gardens are well looked after; they produce all the known European vegetables, and a good many tropical plants besides.

Extensive *patios*, or courts, surrounded by overhanging galleries, serve as shelter to the large caravans which descend from the interior to the coast for the purposes of traffic, and remain three days under cover in order to effect the barter.

A river, which, in the summer season, looks scarcely more than a broad ribbon of white sand running from the mountains to the sea through the forest *do Cavaco*, constitutes nevertheless the great source or spring of Benguella, whose wells, that have been dug there, produce excellent water purified in its passage through calcareous sand.

The broad and straight streets of the town are planted with two rows of trees, for the most part sycamores, but of no great age, and as yet therefore somewhat small. The squares or *places* are of vast size, and in a public garden are flourishing many fine plants that are very agreeable to the eye.

The houses, which have no upper story, are built of unbaked bricks, and the flooring is composed either of tiles or wood.

The custom-house is a good building, recently erected, and has spacious warehouses for the storing of goods. This establishment and the public garden before alluded to, as well as other improvements in Benguella, were the work of a former governor, Leite Mendes. To him also is due, I believe, the foundation of a magnificent pier with iron architraves, subsequently carried to completion by Governor Teixeira da Silva. It is furnished with two cranes and trams, by which goods are conveyed from the vessels into the custom-house. I am grammatically wrong, however, in using the present tense in respect of such conveyance; I should rather employ the conditional, and say they *would be conveyed*, if there were any men to do the work; but as these are wanting, they are not conveyed at all.

The town further boasts of a decent church and a cemetery, well placed and walled in.

The European population is surrounded on all sides by *senzalas*, or the huts of the negroes, which in fact are occasionally discoverable in deserted grounds in the very midst of the dwellings of the whites. Take it for all in all, the general aspect of the place is agreeable and picturesque.

Benguella has a somewhat doubtful reputation among the Portuguese possessions in Africa. Many suppose the country to be infected; that it exhales pestiferous miasma too often causing death from plague. But this is really not the case. True, I was not acquainted with the Benguella of the past, but I can aver that at the present day it is neither better nor worse than many other places in Africa.

Cleanliness and plantations of trees must certainly have considerably modified its former hygienic conditions, and a small amount of goodwill would make it, sanitarily, far better than it is. This cannot fail to be done as time goes on, inasmuch as it is not likely that a place of so much importance, from a commercial point of view, and which is in such close contact with the rich lands in the interior, can remain neglected.

The chief products which make up the trade of Benguella are wax, ivory, india-rubber, and orchilla weed, which are conveyed to the town by the caravans from the interior. These caravans are of two kinds. Some, under the guidance of agents of the trading houses, carry back to the firms which despatch them the products of their trade with the interior; others, composed exclusively of natives, come over to trade on their own account, as being more profitable to themselves.

The trade with the natives is effected by direct exchange of their produce for cotton stuffs, white, striped, or printed. Other European products form the object of a second exchange for the stuffs already received; and thus, after the first barter of the ivory or wax for cotton, the latter is given for arms, powder, rum, beads, &c., at the will of the buyer, because cotton stuffs are, so to speak, the current money of this traffic.

The trade is in the hands of Europeans and creoles, and we fell in there, fortunately, with a good many of those adventurous young spirits who leave their homes and country to seek for fortune in these distant climes.

A few convicts of minor importance also do some trade, either on their own account or as the employés of foreign houses.

The greatest of the criminals of the mother coun-

try—those for instance who are transported for life—
are sent to Benguella, and as a natural consequence a
good number of rascals are to be met with there, to
whom it is well to give a wide berth; taking care not
to confound them with the many really honest and
worthy folks who occupy the place.

The police duties are entrusted to a military force
told off for Benguella from one of the regiments, and
from Benguella itself various forces are scattered
among the communes of the interior, thus weakening
the garrison of the town, which is small enough, in all
conscience, already.

We possess two armies, one in the mother country,
the other in the colonies, which have no connection
between them.

Our home army is good, because the Portuguese are
good soldiers; our colonial army is bad, because the
blacks, of which it is composed, are bad soldiers, and
the few whites that are mixed up with them are even
worse than the negroes. Transported for offences
which exclude them from society and cause them to
forfeit in Europe the rights of citizenship, they follow
in Africa the noble calling of a soldier, by which it
happens that our African autonomy and the public
and private safety are entrusted to the defence of men
who can give as sole guarantee a past career of crime
or misdemeanour.

Hence the constant scenes of a shameful character
that are there enacted. During my stay in Benguella
an impudent burglary was committed in the military
department, and a large sum of money was carried off.
The Governor displayed extraordinary energy in his
endeavours to discover the thieves, and received great
assistance at the hands of his secretary, Captain Barata;
and in the end their efforts were successful, both in
catching the rascals and recovering the money. It will

scarcely be credited that the robbery was planned by the very sergeant of the detachment, and was carried out by him with the aid of some of the soldiers!

If our army at home can escape the censure of fastidious military critics, our colonial forces are objects for the well-merited lash of all foreigners who deign to bestow upon them any attention.

The more I consider the matter, the more puzzled am I to explain the *raison d'être* of such an army as we possess in the colonies, which is neither of use as a police force, nor for the purposes of war; nay, as regards the latter, I remember to have seen better work done by a corps of volunteers raised within the kingdom, and who, besides, were bound by a fixed term of service, than by any of the so-called regulars in the colonies. Even at the present time in Lisbon there are three battalions always ready to start for the colonies, and who have in fact already been there; a proof, in my opinion, that the keeping up an army abroad, on its present footing, answers no other purpose than that of perpetuating a bygone usage.

On the night of my arrival at Benguella I made the acquaintance of the Judge Snr. Caldeira, who was good enough to join the Governor in assuring me that he would use all his influence to prevent my visit to Benguella being abortive, and he kept his word.

The Governor called a meeting, at his own residence, of the most important inhabitants of the town, and, explaining to them the motives of my journey and its proposed direction, begged them to render me every assistance in their power in the way of procuring me carriers, and thus enable me to carry out my mission. This they all promised to do.

H. E. Snr. Pereira de Mello and the judge were indefatigable from this moment, so that on the 17th inst., the day on which the latter left for Lisbon, I had got together the number of men I asked for, viz. fifty, which, with the thirty expected from Novo Redondo, made a total of eighty; as many as I deemed necessary for the journey from the mouth of the Cunene to the Bihé.

The old settler Silva Porto undertook to convey to Bihé the heavier portion of the baggage, which we could take up at that place, and where we should have to engage fresh carriers to pursue our journey.

On that day I shifted my quarters to the house previously occupied by the judge, although I continued to dine with the Governor and occasionally with Antonio Ferreira Marques, of the firm of Ferreira and Gonçalves, who vied with each other in their polite attentions to me.

Next morning a black in my service robbed me of some 75 milreis, and disappeared without leaving a trace of the road he took.

On the 19th my companions arrived on board the gunboat *Tamega*, and on the same day we resolved that we would not go to the mouth of the Cunene, but make our way directly to the Bihé.

This fresh resolution altered the engagements we had taken with the carriers, and besides this, the people of Benguella, who, when led into a distant country, would not think of deserting, might perhaps feel inclined to do so when journeying at the outset through territory whose language and customs they were acquainted with.

And so we had again to alter the plan of our campaign. I kept constantly in my mind the narratives of Cameron and Stanley in respect of the trouble and annoyance caused by desertions, from which indeed not even Livingstone was free, seeing that he was abandoned by thirty men on his Téte journey with Dr. Kirk.

Immediately after the arrival of my two companions it was determined that Ivens should have the charge of the geographical department; that Capello should devote himself to meteorology and natural sciences, and that I should attend to the auxiliary staff of the expedition, whilst giving each other, of course, mutual advice and assistance. As my duties therefore compelled me to set things going, I began by taking counsel of Silva Porto.

I recounted to him the fresh determination we had come to, viz. to proceed directly to the Bihé, and explained to him the difficulty in which I was placed. Silva Porto came over to Benguella with me, as his house—Bemposta—was some four miles distant from the town, and called at the various houses where caravans of Bailundos might be found, without however succeeding in getting any offers to carry the baggage to Bihé. We then learned that a large caravan had arrived at the house of Mr. Cauchoix, and proceeded thither; that gentleman did his best to help us, but in vain, although he offered a heavy gratuity to the chief, and double pay to the porters, if they would only take our things.

I wish to mention here a very curious fact. The Bihenos are the finest travellers in Africa, and no other people extend their journeys to such length as they, or can equal them in pluck and endurance under fatigue; but these Bihenos only travel from Bihé into the interior as hired attendants, for if, which is very rare, they come down to the coast, it is on their own account. The Bailundos, on the other hand, hire out their services between the coast and Bihé, and will not go into the interior in an easterly direction; northwards, however, they have no objection to extend their journeys to the Dumbo and Loanda.

Thus it happens that merchants settled in the country

have their goods transported from Benguella to the Bihé country by Bailundos, and thence to more remote places in the interior by Bihenos, who come back to Bihé laden with products in exchange; and from Bihé to the coast, the Bailundos resume the service.

Having obtained this information, all that was left me to do was to hire some Bailundos to come over and fetch the baggage; and Silva Porto having kindly undertaken the task of procuring them, despatched at once five blacks to Bailundo for the purpose. The old trader, however, did not fail to assure me, from his long experience, that a good deal of delay must be expected, as it would take his messengers fifteen days to reach the country, and at least as many more to collect the carriers; so that, adding these thirty to fifteen others for the return journey, we must reckon upon forty-five days ere they got back; and there was little chance indeed of their being here before. We were then at the end of September, so that by this computation we should not be able to start before the middle of November.*

After taking counsel with my friends upon this fresh phase in our position, we resolved not to lose such valuable time at Benguella; but, delivering over the heavy baggage to Silva Porto, for him to forward it by the Bailundos, start at once with such things as were indispensable, and wait for the remainder at Bihé. The time we spent there could at least be occupied in hunting up fresh carriers to pursue the onward journey.

Out of the men hired at Benguella we could not reckon with confidence on more than thirty performing the journey, and these, with thirty-six obtained from Novo Redondo, made a total of sixty-six men. Besides

---

* As a matter of fact, a portion of these porters, viz. 200, only reached Benguella on the 27th of December, and 200 more at the end of February.

these, we had fourteen soldiers, some young niggers for my personal service, two or three Kabendas in the service of Capello and Ivens, and two native chiefs, one of whom, Barros, had been engaged by me in Katambela, and the other, Catão, by Capello, in Novo Redondo.

Among all these men, there was not one in whom we could repose any confidence.

We set to work to select the loads judged indispensable, and found that they were eighty-seven, thus making twenty-seven more loads than there were carriers. No one can conceive how I laboured to supply the deficiency; but in vain, not another porter was to be had.

The blacks, not understanding what we intended to do in the interior, became uncomfortable, and, with their naturally suspicious nature, got all kinds of fancies, which did not improve matters.

The end of October came, and still we were in the same position.

By the advice of Silva Porto, I made up my mind to go into the Dombe country, and see whether the Mundumbes would be more difficult to deal with than the people of Benguella. Feeling, however, indisposed, I got Capello to go for me.

Capello started on the 29th and returned on the 3rd of November, having made a fruitless journey. The Mundumbes are willing enough to go to Quillengues by a road known to themselves, but beyond this they will not travel, and they refused the very handsome sum we offered if they would go with us to Bihé.

It became absolutely necessary to come to some determination, and that we therefore at once proceeded to do. We resolved still to go to Bihé, but by the track leading through Quillengues and Caconda.

The Governor, Pereira de Mello, immediately gave

orders to the *chefe* (head official) of the Dombe to have ready fifty carriers, to accompany us to Quillengues; and Silva Porto, as agreed, took charge of the baggage which was to be forwarded to Bihé, amounting in all to 400 loads.

His Excellency placed at our disposal a large boat to convey by sea to the Cúio (Dombe Grande) the loads that had to be transported thence to Quillengues, and certain of the Benguella carriers who were on the sick list.

On the 11th of November we were ready to leave the coast, and we fixed our departure for the following day. On the former date four of the Novo Redondo porters ran away, and five of the Benguella on the very morning of our departure.

The 12th arrived at last, and with it our final exit from the town, after the most cordial adieux and good wishes of the many friends assembled to wish us God-speed. Shortly before leaving I had gone down to the beach and feasted my eyes on the vast expanse of the Atlantic, on that enormous waste of water which I gazed on, perhaps, for the last time. Two years did indeed elapse before I had the satisfaction of seeing it again, and then it was in France, near Bordeaux.

I do not know if all persons are affected in the same way, but, after I have dwelt for any time in a place, I quit it with regret. At the moment of leaving Benguella, I felt a pang of sorrow, an indefinable sensation of *malaise*, which I must confess the town and its surroundings could scarcely of themselves be held capable of exciting.

The national colours, carried by one of our party, were increasing their distance from the town, as our caravan wound its measured way into the open, and with one more hasty farewell I hurried after it.

On the 13th we reached the Dombe, having made a journey of 40 miles. We had with us sixty-nine persons and six donkeys, which were all, men and asses alike, lodged in the fortress. We three, with our body servants, were most kindly welcomed to the house of Manuel Antonio de Santos Reis, a perfect gentleman, who could scarcely do enough to serve us.

It was a couple of days later that our baggage, which had been sent by sea, arrived, and after a careful examination of the whole I found 100 men, besides those I had with me, would be necessary for its transport.

This arose, I presume, from an abuse of the accomodation offered us by the boat, more things being put on board than those we at first judged absolutely necessary.

We decided upon leaving on the 18th, after receiving our letters from Europe, as the packet usually reached Benguella on the 14th; but not only on the 18th had the steamer not arrived, but the *chefe* had not hired a single porter.

The mail came in on the 21st, but as regards followers we still had only those we brought from Benguella. The *chefe* declared all should be ready by the 26th; but so far from this being the case, only nineteen out of the hundred required appeared on that day. Next morning we procured twenty-seven more; when, fearing if there were any greater delay those I had already obtained would take themselves off, I at once despatched them to Quillengues, under the charge of two of the soldiers I had with me.

The *chefe* asseverated that it was impossible for him to get any more men. Whereupon I invited to the fortress the three Sovas (native chiefs or princes) of the Dombe for the 28th, in order to see whether I could not myself treat with them. They came—three magnificent specimens, whose appearance was calculated to strike a beholder with surprise, if not with awe.

One was called Brito, a name he had borrowed from a former Governor of Benguella, who had restored him to power; the second, Bahita; and the third, Batára. My companions unfortunately could not be present at this serio-comic meeting, as they had been suffering since the 24th from fever.

Sova Brito was attired in three petticoats of chintz, of a large flowered pattern, very rumpled and dirty, with an infantry captain's coat, unbuttoned, displaying his naked breast, for shirt he had none; and on his head, over a red woollen nightcap, was jauntily posed the cocked hat of a staff officer.

Bahita also wore petticoats, of some woollen stuff of brilliant colours, a rich uniform of a peer of Portugal, nearly new, and on his head, over the indispensable nightcap, a kepi of the 5th Chasseurs.

As to Batára, he was dressed simply in rags, but had buckled about his waist an enormous sabre.

These illustrious and grave personages were surrounded by the satellites and high dignitaries of their negro courts, who squatted on the ground about the chairs on which their respective sovereigns were seated. Bahita was accompanied by a minstrel who played upon a *Marimba*, from which he drew the most lugubrious sounds.

This instrument is formed of two sticks about three feet in length, slightly curved, there being stretched from end to end strings of catgut on which are fixed thin strips of wood, each of which is a note of a scale. The sound is increased by means of a row of gourds placed below, so arranged that the lowest note corresponds to a gourd having a capacity of six to seven pints and the highest to one of a quarter of a pint or less.

The sovas conducted themselves with such extraordinary gravity that in spite of myself I imitated their example.

After having promised me carriers, they were good enough to accompany me to my temporary home, about a mile and a half from the fortress; and as I made each of them a present of a bottle of *aguardente*, they ordered their chief officials to honour me with a dance, and Bahita commanded some girls, who had hitherto been kept out of sight, to be brought forward to join in the entertainment.

I begged them to dance themselves, but they gave me to know that their dignity would not allow of such a proceeding, it being contrary to all established rule. I ardently desired, however, to see Bahita capering in petticoats and a peer's uniform; and aware of the power of liquor over the negro, I gave instructions that a fresh bottle should be presented to their majesties.

This was quite enough. Laws and established rules were soon cast to the winds, and I had the delight to see them all join in a grotesque dance in the midst of their people, who, fired with enthusiasm at the sight, rolled about and went through such violent contortions that one would have thought they had all gone into fits or were afflicted with some new kind of madness. Bahita was simply grand, and I cannot help thinking that the "roi Bobèche" must have been created after some such model. In his excitement he talked of nothing but ordering people's heads to be cut off; sentences which those around him listened to with the utmost apparent submission, with their tongues in their cheeks all the time, as they knew full well the Portuguese Government would allow of no tricks of that kind within its jurisdiction.

The Dombe Grande is a most fertile valley, which extends first from south to north, and then westwards, almost in a right angle, to the sea. It is framed in by two systems of mountains, one on the west, which borders the coast, and the other on the east; and through it runs a river known under no fewer than four names, the *Dombe, Coporolo, Quiporolo,* and *St. Francisco.*

This river, very full of water in winter, is generally quite dry in summer, although, even in the times of greatest drought, water can always be had by digging wells; this is the case, in fact, throughout the Dombe valley, where one never need go deeper than ten

Fig. 1.—Mundombe Women, Vendors of Coal.
(From a photograph by the chemist Monteiro.)

feet to obtain the desired supply. Close to the western mountains, in that part of the valley which runs north and south, there is a lake, fifty-four yards wide by five-eighths of a mile in length, of the shape of the letter S. This lake is curious, inasmuch as it is not formed by rain deposits, but is fed by a strong subterranean spring: its level is never changed, the surplus being carried off by infiltrations which, less than a mile lower down, jut out in the shape of springs, that are made use of for irrigating some property in the neighbourhood. The lake is said to contain some large fish and many crocodiles.

I visited it frequently, but never caught sight of either crocodile or fish. I must believe, however, that they exist, because my kind entertainer assured me of the fact, and that they were very voracious to boot. He stated, besides, in corroboration of his assertion, that in 1876, his place having been attacked by a band of marauders from Quillengues, the latter were defeated by his blacks, and attempted, in their flight, to swim the lake. Not one, however, reached the opposite bank, the whole of them having fallen a prey to the voracious denizens of the waters.

In those same western mountains, which are formed of calcareous carbonate and some sulphate of lime, and in close proximity to the lake, exist certain huge grottoes or caverns, which, as we were informed by our host, had never been explored, and which contained, in so far as could be observed from outward inspection, extensive galleries.

Capello, myself, and our host, Snr. Reis, went to visit one of them, and found that it had been greatly exaggerated.

It formed a species of hall, nearly circular, of about 15 yards in diameter, scooped by nature out of the immense mass of calcareous stone of which the mountain was composed. It would seem to be a regular haunt of wild beasts, as one might judge from the air, which was perfectly saturated with the pungent smell of certain animals, as well as from the traces of a lion impressed on the impalpable powder which covered the ground, where we met with a few quills of the Hystrix Africano.

In the valley of the Dombe there are some important agricultural estates, the chief of them being that of the

Loache, one of Paula Barboza, and that of our host, Santos Reis. The last mentioned is scarcely three years old, and produces sugar-cane in sufficient quantity to yield more than eight thousand gallons of rum; and it must be remembered that the land was previously all forest-grown, and has only been three years cleared. The estate is otherwise still in its infancy, everything being in course of construction; but one may readily

Fig. 2.—MUNDOMBE WOMEN AND GIRLS.
(From a photograph by Monteiro.)

judge, from the results already obtained, how richly productive is the soil in this part of the world.

The entire valley is cultivated with manioc by the natives, and is so fertile that even after three years drought its production is perfectly regular, more than fifty thousand bushels of the flour being exported during the year. It is, in fact, the granary of Benguella. The natives of those parts do not trade by barter, but sell their products for money, the value of which they are very well acquainted with.

Our compulsory delay in this country was most injurious to the order and discipline of my people.

Every day they put forward some fresh claim; every day some quarrel or other arose among them; and I feared to be too strict lest they should all desert me in a body.

They sold their clothes to purchase aguardente, and even went so far as to dispose of their rations of food to procure liquor wherewith to muddle themselves.

Fig. 3.—Mundombe Men. (From a photograph by Monteiro.)

The soldiers were the worst. The Sovas did not send us any men, and I began to apprehend a repetition of the Benguella scenes—any way we could not stir.

On the 1st of December thirty men arrived at Dombe, sent from Quillengues by the military *chefe*, to fetch some baggage belonging to him. I at once pounced upon them, and arranged with my companions to start on the 4th.

We had to record three other desertions: two men from Novo Redondo, and one from Benguella.

Our donkeys were very troublesome and obstinate, and there was no one who knew how to train them; the parting with them was, however, out of the question, so we managed as best we could.

CHAPTER III.

THE STORY OF A SHEEP.

Nine days in the desert—Want of water—The *ex-chefe* of Quillengues—I lose myself in the bush—Two shots in time—A little nigger and a negress missing—Loss of a donkey—Quillengues at last—Death of the sheep.

ON the 4th of December I left the Dombe, at eight o'clock in the morning, and bent my course to Quillengues. Capello and Ivens remained behind for a while to arrange about sending on some of the luggage, intending to join me at night. By the advice of the guides, we did not follow the caravan route, but a by-path known to themselves, so as to avoid the usual fords of the River Coporolo, which were already somewhat difficult on account of the quantity of water, whilst the other path led to shorter and more convenient fording-places.

After two hours' march in the plain, we arrived at the foot of the Cangemba range, which borders the valley of the Dombe on the east side. Here we got a little rest, and at eleven started off again, endeavouring to cross the mountain by the bed of a torrent, then dry. It was difficult work. The men were heavily laden, for, besides the actual loads of the expedition, weighing 66 pounds, they carried rations for nine days, in the shape of manioc flour and dried fish. The difference of level was barely 550 yards; but the bed of the torrent, formed of calcareous rock, offered formidable obstacles to our progress. In many places it was necessary to use our hands as well as feet to get along, and the getting the donkeys over the ground was a work of considerable difficulty.

We had purchased in the Dombe a couple of sheep, to be killed upon the road, and one of them followed our party readily enough; the other, however, caused us a good deal of trouble, by not only refusing to follow but showing a great and constant inclination to return to the country we had left behind.

Three hours were spent upon our fatiguing march, and in covering a thousand yards at most of ground. The sun poured down upon us as we toiled on, unsheltered, and we were fagged out with our exertions. We encamped at length beside a well dug in the sandy bed of a rivulet that had run dry, and to which little stream the Mundombes gave the name of Cabindondo. The spot was an arid one, and only here and there were visible some white thorns, curled and burnt by the sun, which at this period of the year literally pierce like a knife. Our horizon was formed by the summits of the mountains which run north and south.

Towards evening Capello and Ivens put in an appearance, and we at once sat down to our meal—not before we needed it, and I, indeed, was still fasting. On the 5th, at early morning, we were on the move in a S.E. direction, and after four hours' march, during which we got over a space of twelve miles, we pitched our tents in a place which the guides called Taramanjamba, an extensive valley, surrounded by hills of no great height. The altitude was found to be 656 yards, thus showing that we were scarcely more than 110 yards above our camp of yesterday.

Vegetation continued poor, and the want of water was great. For drinking and cooking purposes we obtained but little, in the shape of rain deposits in the cavities of the rocks; deposits immediately exhausted by our thirsty caravan, so that as night came on thirst was sensibly felt.

During our march, if the young asses continued to be troublesome, the sheep above referred to was no less so: he was wonderfully wild, and more obstinate than the donkeys. I determined to have done with him, and my companions being of the same mind I gave orders to the niggers to this effect, and took a stroll in the environs.

On my return to the camp, I discovered that the stupid fellows had misunderstood my orders, and instead of killing the wild sheep had made away with the quiet one.

On the following morning we started at daybreak, and after five hours' march pitched our camp at a placed called Tiue, where our guides assured us we should find water.

Against all expectation, the sheep whose life had been saved by accident not only gave over his wild tricks, but took it into his head to follow me about like a dog, keeping constantly by me, whether on the march or in camp.

The journey was a difficult one that day; for my people were parched with thirst, and for upwards of an hour we had to follow the dry bed of the river Canga, naturally all stones and irregularities, which fatigued us very much.

The soil is granitic, and the arborary vegetation luxuriant.

The water, just as the night before, was rain water, collected in the cavities of the rocks; but it was more agreeable to the palate, and clearer to the eye.

Some of our men had wounded feet, so that it was dark ere they reached the camp, as they could only crawl along; there were others who followed their example out of weakness, and many more from sheer sloth.

On that day, among the laggards were unluckily the carriers of the commissariat, which made it late before we got any food. Capello, quiet and undemonstrative, never complained of the inconveniences he was put to; but he was silent under them. Ivens, on the contrary, was always full of spirits, and with his loquacity and light-heartedness kept us in good-humour, and often made us merry with his witticisms. His appetite, which was never at fault, was great on this occasion, and after the arrival of the carriers he watched with eager eyes a leg of mutton which a nigger was turning before the fire on a wooden spit. At last he exclaimed: " If my father could only see me eyeing that joint, I am sure the old man would be moved to tears!"

Since leaving the Dombe we had scarcely eaten once a day, which was the case also with our people; with this difference, however, that they ate without interruption from the moment of camping until they went to sleep; which made me, not unnaturally, apprehensive that the rations given out for nine days would be very soon exhausted, and that hunger would follow in a country where it was impossible to obtain food.

On the following day we made sixteen miles in an E.S.E. direction, and pitched our tents in a forest called Chalussinga; the ground, still granitic in character, was relatively better walking, and the vegetation was of a more vigorous kind than we had hitherto seen.

We met in this forest with the first baobabs we had seen since leaving the coast. Water continued to be scarce, and was always formed of rain deposits. At about three in the afternoon of that day we were advised that a caravan was coming in the direction of our camp, on its way from the interior; and on issuing out to meet it, we found that it was the ex-chefe of

Quillengues, Captain Roza, on his way to Benguella in ill-health. We invited him to our tent, where he dined, and at parting we were able to furnish him with some medicines, of which he stood greatly in need.

After he had left I was informed by the young niggers that round the camp there were fresh tracks of game, and I went out to investigate. I followed the trail of some large antelopes, and it led me so far that night fell, with a darkness so profound that I lost all traces of the way back to the camp. A lofty mountain stood out in sombre relief against a hazy sky, where not a single star was seen to glitter. It occurred to me to scale it, so that I might from some elevated pinnacle discover the lights of the camp, by which to direct my steps. I deemed the notion a happy one, for having ascended the mountain I discovered in the distance a gleam of light, which I at once made for, having marked the direction by my pocket compass.

None but those who have experienced it can imagine what it is to wander on a dark night through the brambles and underwood of a virgin forest, and how much time is expended in traversing a brief space, leaving, by the way, here a fragment of clothing, and there, it may be, a portion of one's skin.

I arrived at length, guided during the latter part of the route by human voices; but judge of my surprise and disappointment at finding that I had mistaken Captain Roza's camp for my own, and that I must still be some four miles distant from the latter! As however a road, or rather the track left by a caravan, connected the two camps, I determined to push on, by its guidance, and after another hour's tramp I heard the welcome sound of the horns blown by my people, and the occasional crack of a rifle fired off to attract my attention and direct my steps.

I reached my tent completely tired out and wounded

with the thorns, and found Capello and Ivens in no little anxiety on my account. Nor was I allowed to go undisturbed in mind to the rest I so much needed, for I was informed, to my annoyance though not to my surprise, that provisions were falling short, and that the soldiers especially had in five days consumed the rations of nine.

We made a somewhat forced march next day, and in six hours covered 18 miles, still travelling E.S.E.

The road was a good one, as we followed the track of Captain Roza's caravan. Gigantic baobabs continually appeared in the forests we passed through. It was after crossing the river Calucula that we pitched our tents, selecting a spot on the right bank of the stream.

The river boasts of but little water; but what it contains is limpid and good.

We still continued eating but once a day, the hour for the single repast varying from one to three, according to the journey. It had become necessary to be parsimonious with our stores. I still felt the fatigues of the previous night, and therefore remained within the encampment, instead of hunting up game. Ivens, as usual, employed himself at his drawing; and Capello was busy with his collection of insects and reptiles.

The soldiers, having finished their rations, began to complain of hunger, and even talked of killing the sheep. I had taken quite a liking for the animal, which had been so suddenly converted from the wild creature it was into a gentle and domestic beast, following me, as I have mentioned, constantly about, and never allowing me out of its sight. The idea therefore of killing it was very repugnant to me, and Ivens for the time diverted the soldiers' attention by giving them a little rice from our own stores.

On the 9th we broke up our camp at five in the morning, and kept steadily on our march till one, when we rested on a slope of Mount Tama. From eight till nine we travelled southwards by the left bank of the river Chicúli Diengui, which runs north, probably into the Coporolo. Vegetation was becoming more and more luxuriant, and on this day our route lay through a dense forest.

Directly our tents were pitched, the complaints of the hungry soldiers were again very audible, and the subject of killing the poor sheep was once more mooted. Ivens gave the fellows another ration of rice, which satisfied them for the time; but of course it was only staving off, as it were, an evil day, and could not be considered as a positive salvation for the poor animal.

Fagged out as I was, I resolved to go hunting for game, with a view to save the life of my poor sheep.

For upwards of an hour I rambled through the forest without result, and was turning my steps campward when, in a small open space of ground, I sighted two antelopes grazing.

I drew near, but at more than a hundred yards distance my presence was evidently discovered. The male leaped upon a rock, and there began to cast his keen eye in every direction, whilst the female, with ear on the alert, sniffed about her.

The distance was great, but I did not hesitate to fire, aiming at the male, which I had the satisfaction to see fall and roll over. His companion, hearing the report, sprang on to the rocky ground, when I discharged my second barrel. With one bound, however, she then disappeared in the underwood.

My young nigger started off to secure the dead antelope, but I perceived that instead of stopping at the rock where the creature was last seen, he turned aside and went farther on, and I myself at length arrived at the spot, and began, with an anxious feeling at the heart, to search all round, for I feared I was mistaken in seeing the first antelope fall. It was not so, however, for on the other side of the rock, to my great joy, I discovered the graceful animal (*Cervicapra bohor*), stone dead.

I had scarce time to satisfy myself on the point than my attendant appeared from the wood bending under a heavy burden.

It was the second antelope, which he had found dead at no great distance from the open ground. Both animals had been struck in the breast, but whereas the male had been killed upon the spot, the female had made a bound or two before she ultimately fell.

The sheep, then, was for the time saved, and indeed, as in two days' time we ought to reach Quillengues, where provisions could be had, the poor beast might be looked upon as perfectly secure.

On the following day, after a march of twenty-two miles, and wading across the rivers Umpuro, Cumbambi, and Comooloena, we encamped on the right bank of the Vambo; all four streams run northwards, to unite their waters—when they have any—to those of the Coporolo, here already called the Calunga, a name which it retains up to its source.

During this day's march we fell in for the first time with enormous grasses, clothing the open spots of the wood. So tall and thick were they that it was quite impossible to see over them, and very difficult to effect a passage through. In the course of the journey one of my young niggers disappeared, together with a negress, the wife of Capello's attendant Catraio; and though I sent out people to look for them, they were nowhere to be found.

The scarcity of provisions was great, and it was not the soldiers only who complained of hunger; the whole lot were grumbling and would not listen to reason. There was no help for it—on we must go.

On the 11th, after passing two small streams which the rains had considerably swollen, the Quitaqui and the Massonge, we encamped on the right bank of the river Tui, very near to Quillengues. There was no news of the missing youth and woman, and since the evening before one of the asses had disappeared. Whilst the men were busy with the camp, I started off for the fortress of Quillengues, in search of stores, with which I returned at eight in the evening. Decidedly the sheep was saved.

During the night the young negro and negress, whom we thought lost, found their way into camp, a circumstance which gave me much pleasure; for, compelled as we were, by hunger, to go on, we could not have lingered to search for them.

The place where we had pitched our tents was low and marshy, without any conveniences at hand, and isolated. On this account, we resolved to shift our quarters and encamp in the compound of the *chefe* of Quillengues, which we reached at eleven o'clock in the forenoon of the 12th of December.

I there paid and discharged the carriers from Dombe who had engaged to come with us to Quillengues, and I begged the *chefe*, Lieutenant Roza, to obtain others for me to Caconda. This, he assured me, would be easy; only, as he was informed that the streams between Quillengues and Caconda were too full to allow of crossing, we should not be able to start immediately.

We fed well on that day, in fact we had two meals, breakfast and dinner.

A few days after this, the donkey which had been lost in the woods was brought into camp by a native, who had found it strolling about. I gave the negro a gratuity to encourage him in his honesty; and besides, I never expected to see the poor animal again, for if it escaped the teeth and claws of the wild beasts, it could not, as I thought, avoid capture by some wandering thief.

Quillengues is a valley watered by the Calunga (a river which I suppose to be the upper course of the Coporolo), is extremely fertile, and covered with a native population.

The Portuguese establishment occupies an area of 56,875 square yards; it being of rectangular shape, 273 yards by 199. This rectangle, surrounded by a palisade, has four bastions, built of masonry half way up each face; and within are barracks which form the residence of the military *chefe* and quarters for the soldiers.

Some baobab-trees and sycamores shade with their gigantic branches and thick foliage a ground covered with the huge native grass which affords pasture to the *chefe's* flocks.

If the importance of Quillengues is great as a productive centre, and easy of colonisation, it is not less so as a strategic position; inasmuch as it may be considered one of the keys to the interior, with respect to Benguella.

The petty chiefs of the country acknowledge the Portuguese authority; but being by nature predatory, they attack unceasingly other native tribes, and carry off their cattle.

They are more pastoral than agricultural, but notwithstanding cultivate the land, which yields abundantly to the slightest care, producing maize, massambala, and mandioca or manioc in large quantities.

Their dwellings are circular huts from ten to fifteen feet in diameter, constructed of the trunks of trees, plastered with mud. The door is sufficiently high to afford passage to a man without stooping.

The inhabitants of Quillengues are tall of stature

and robust, and are by nature bold and warlike. Their manufacturing powers are not remarkable, and do not seem to go much further than the fashioning out of iron their assegais, arrow-heads, and hatchets, both for warlike purposes and cutting wood. Their metal castings are not made at home, but are purchased in the Dombe country or in Benguella.

Their folds, like their villages, are surrounded by a strong palisade: which is further protected, exteriorly, by a thorny abattis, to guard against the night attacks of wild beasts.

The mandioca fields are similarly protected by thorns, for small deer (*Cephalophus mergens*) abound, and, from their extreme liking for the leaves of the mandioca, cause serious damage to the plantations.

Aguardente is in great favour with the Quillengues, and so given are these people to drunkenness, that during three months in the year—so long in fact as lasts the fruit of the *gongo*, from which a fermented liquor is made—they are constantly in a state of intoxication, and no possible service can be got out of them for love or money.

When a man is desirous of matrimony, he sends to the father of the lady of his choice a present, which must be composed of four yards at least of cloth from the coast, and a couple of bottles of *aguardente*. The bride comes back with the bearer of the gift, accompanied by her relatives, when a great feast is held, whereof the *pièce de résistance* is an ox offered by the bridegroom.

Adultery is held in high favour by husbands in this part of the world, as their barbarous law enables them to get a heavy fine out of the lover in the shape of cattle and *aguardente*. A wife who has no peccadilloes to answer for gains but little favour in the eyes of her lord, as she does not help to increase his store.

When the lady has fallen off from her duty, she goes to her husband to complain of having been led astray, and upon the accusation of the wife a conviction is obtained.

When a death occurs, the body is shrouded in a white cloth, and being covered with an ox-hide is carried to the grave, dug in a place selected for the purpose. The days following on an interment are days of high festival in the hut of the deceased. The native kings are buried with some ceremony, and their bodies, being arrayed in their best clothes, are conveyed to the tomb in a dressed hide. There is great feasting on these occasions, and an enormous sacrifice of cattle, for the heir of the deceased is bound to sacrifice his whole herd in order to regale his people and give peace to the soul of the departed.

On the 22nd we had a disastrous event occur in our camp.

One of my young negroes stole a Pertuisset explosive bullet, and in company of two of his fellows resolved to let it off, in order that each might have a piece of the lead. Resting the bullet on a stone, one of them placed a knife across it, which he struck with a violent blow, the other two standing near to watch the sport. The bullet suddenly exploded, wounding all three, one of them—by name Silva Porto Calomo—severely, as he received in different parts of his body thirteen fragments of the desired lead, many of them producing deep wounds.

We sent off some men to reconnoitre whether the rivers were yet fordable, and learned from them that the waters were still high—not a very surprising fact, as it had not left off raining during the whole time of our encampment. We thereupon resolved to take another road, which, though considerably longer, was not incommoded with water, and in consequence begged the *chefe* to have some carriers ready. This he did, and on the following day I allotted the men their loads. I felt, however, so extremely poorly myself, that though I sent the porters on I was obliged to stop behind, my friends remaining with me to bear me company. I struggled against a violent fever for three whole days, and was quite unconscious during the 25th, Christmas Day, and the anniversary of my daughter's birth.

I was carefully nursed by Capello and Ivens, the *Chefe* Roza, and his wife, and on the 28th was able to rise from my bed and go out. It was then determined that we should leave on the 1st of January 1878, that is to say, three days afterwards.

The wife of Lieutenant Roza made me two presents, which I was far from thinking would play an important part, later on, in my journey. They consisted of a Sèvres tea service and a remarkably tame she-goat of small breed, on which I bestowed the name of Cora.

Just at this time occurred a disaster which caused me sincere regret. My poor sheep, on whose behalf I had willingly borne so many annoyances with my hungry followers, was killed through a setter that I had brought with me from Portugal and had made a present of to Capello. Pursued by the dog, it endeavoured to force its way through an opening in the palisade, and broke its leg and otherwise injured itself, so that it shortly died. It was my first great trouble during a journey so fruitful in mishaps.

CHAPTER IV.

THROUGH SUBJUGATED TERRITORY.

Journey to Ngóla—The native king Chimbarandongo—Beauty of the country—Arrival at Caconda—José d'Anchieta—No correspondence—Arrival of the *chefe*—We follow the carriers—Ivens goes to the Cunene, and I go to the Cunene—Return from Bandeira's house—Carriers wanting—My opinion.

On the 1st of January 1878 we quitted Quillengues, where we made a good provision of food, and purchased several oxen and sheep, to be slaughtered upon the journey. The *chefe*, Lieutenant Roza, accompanied us a few miles on the road, when he returned to his simple home, and we kept on our course, in a S.E. direction, till we reached the foot of the Quillengues range, where we camped close to the village of Seculu Unguri.

We had on this occasion a travelling companion of the name of Verissimo Gonçalves, who had begged to be allowed to join our party as far as Bihé. He was the son of a well-known Bihé trader who had lately died, and had been acting at Quillengues in the capacity of a clerk to a former servant of his late father's. This young man, a mulatto and but poorly educated, was short in stature and perverted in mind, being full of the vices proper to his race, but was still not wanting in good-nature or intelligence.

I make somewhat particular mention of him, as he will appear again in the course of my narrative.

He was shy and timid, though not cowardly, and

under a rather weakly appearance concealed a strong constitution and muscles of iron. He could scarcely read or write, but was a tolerable shot and a crafty woodsman.

During our stay at Quillengues I had managed to break in two of the asses, which were very useful to me as mounts on this new journey.

The following day, at starting we commenced the ascent of the mountain, here called Mount Quisséeua.

It was excessively toilsome work, and for three weary hours we had to struggle with the asperities of the mountain side, till we reached an elevation of 5700 feet above the level of the sea, or 2740 above the plateau which terminates at Quillengues.

In a defile of the mountain we passed a small rivulet which the natives call *Obaba tenda*, meaning "cold water." We fixed our camp on the bank of another, called *Cuverai*, an affluent of the Qué. These two rivulets are permanent, and their waters flow into the Cunene.

The soil continued granitic, but the vegetation had entirely changed in aspect—due, of course, to the elevation we had reached. The baobab had disappeared, and ferns were nestling in the shade of the numerous and varied acacias which peopled the woods. The flora presented greater wealth of herbaceous plants, and in the grasses more especially the most vigorous vegetation was observable.

I noted that at times we traversed regions where not a single bird was visible, and then, all of a sudden, we would enter tracks where thousands of the feathered tribe almost deafened the ears with their noise. Of larger game there was but little, but there were traces of its existence.

During the night of the following day we had rather a curious adventure. We were encamped beside the

Quicué, a brook running S.E. over a granitic bed, to swell the waters, most probably, of the Qué, when we heard Capello's dog barking furiously at something in the neighbourhood of the hut. At the same time we were conscious of a sound, at no great distance from us, like that of an animal chewing the cud, which induced us to believe that the donkeys had got out and were grazing in the camp that was surrounded by the thorny abattis. We therefore quieted the dog and went off to our beds. Day was just breaking when we heard a great uproar in the camp, and turning out we learned that the blacks, who, at the outset, like ourselves, thought the donkeys had broken loose, had discovered their mistake, and that some strange animal had got into the camp. And so in fact it proved, for an enormous buffalo had done us the honour of a visit during the night.

It was a strange circumstance, and at a first glance difficult of explanation; a clue, however, to the mystery might have been probably discovered in the repeated roarings of the lions, that were plainly audible, and which perchance drove the buffalo to our camp for shelter.

The day after we moved our camp close up to the village of Ngóla, and I at once caused my arrival to be announced to the native chief.

After breakfast I proceeded to the village to call upon him. I was accompanied by my young negro servants, who carried a chair for my use and two parasols.

The chief at once appeared, armed with two clubs and an assegai. He wore a long waist-cloth, and over it a leopard's skin. His chest was bare, and from his neck hung a number of amulets. He received me outside his hut under a burning sun. I offered him one of the parasols I had brought with me, which was covered with thin scarlet cloth, an attention that seemed to please him mightily.

I informed him of the object of my journey, which he did not readily comprehend. He perfectly understood, however, the value of the gifts I made him, and consisting of a small barrel of gunpowder, fifty gun-flints, and a dozen tin grelôt-bells; although my asking nothing in exchange filled him with wonder.

I invited him to my camp to see my companions, to which he agreed, and accompanied me on the spot; a matter worthy of note, as native chiefs are mostly suspicious by nature.

When I told him he might bring a vessel in which to put some *aquardente*, he went and fetched a bottle that would hold about a pint and a half. I could not help being astonished that a chief of his rank should be so little covetous, and desired him to procure a larger vessel. He then sent for a gourd which would contain about a couple of bottles, and I begged him to bring another of the same size. The chief could not conceal his admiration at my generosity.

We set off on foot, accompanied by three of his wives, his daughters, and many of his people, all unarmed, to show me the confidence with which I had inspired him.

We reached the camp at a time when Capello was making meteorological observations, and our guest was lost in admiration at the thermometers and barometers.

Ivens shortly joined us, and after an exchange of compliments showed our noble guest the Snider and Winchester arms, at which he was quite dumfounded.

Chimbarandongo, for that is the name of the native chief of Ngóla, is a man of intelligence, and knows how to make life very tolerable among his subjects.

He offered us an ox, and readily consented to my request to have it slaughtered, as we were in want of provisions. He wished, however, that I would slay it with my own hand.

The ox meanwhile had broken loose, and was making towards the wood. It was already some eighty paces from us, when I seized my rifle, and, telling the chief where I would hit it, fired, and the beast fell.

Chimbarandongo went to examine the animal, and, on seeing the wound, whence the blood was running, just between the eyes, in the very place I had indicated, he was so astonished that he embraced me again and again in his enthusiasm.

At about four o'clock, there broke over us a violent storm, with thunder and lightning and heavy rain, which lasted a couple of hours.

The chief took refuge in our hut with his women folk and a few of his chief followers. He then made them a speech, the object of which was to prove that we had brought down the rain, and with it a vast benefit to the country, then suffering from the excessive heats of summer.

We tried to explain to him that we did not possess any such great powers, and that God only could influence the grand phenomena of nature. It was Ivens who undertook to illustrate how and why the rain fell.

Before the lecture on meteorology was half over, the chief turned his followers out of the hut, and assembling them again at the close of Ivens's discourse, declared that if it left off raining, he would pitch upon the unlucky mortal who was the cause of its ceasing and have him put to death without delay.

Disconcerted at this strange mode of interpreting our well-meant lesson, we addressed ourselves to the task of pointing out to him the inutility of capital punishment, which probably had as little effect upon him as the previous instruction; it was clear, however, that half drunk as he was he had sufficient sense left to discover that our theories harmonised but little with his system of government.

When night fell, his majesty retired in the most comic fashion, mounting pickaback on one of his counsellors, whose hands rested on the hips of another walking before him; and as they were all more or less intoxicated, they reeled about in the most ludicrous way, threatening at every moment to topple over together, and perhaps break the sacred head of their sovereign into the bargain.

This King Chimbarandongo was not wanting in sense or judgment. He did not believe in sorcery, nor did he believe that we had brought down the rain; but it suited him to appear to do so, in order not to lose his prestige among his people, who were quite satisfied with the form of government he imposed upon them.

The next day, when he came to take leave of us, he let me know that it was his policy to remain on good terms with the whites; inasmuch as, through his friendly relations with them, he obtained the cloth which covered him, and the arms and powder that secured him respect from his enemies.

"Without the whites," he said, "we are poorer than the beasts, as they possess the skins we are forced to rob them of; and those blacks are great fools who do not seek to gain the friendship of the palefaces."

The village or hamlet of Ngóla is strongly defended by a double palisade, put up with some art, one of the faces being even so arranged as to allow of a cross-fire. The space inclosed is so vast as to be able to contain the entire population of the country, which gathers there, with all its flocks and herds, when the district is in a state of war. The little stream, called the Cutota, runs right through it, and it is therefore capable of sustaining a long siege without any inconvenience in respect of water.

On leaving Ngóla, we journeyed for a couple of hours in a N.E. direction, till we fell in with the Qué, the largest of the rivers running between Quillengues and Caconda. At the spot where we attempted a crossing we found the stream at least 50 feet wide and from 12 to 16 feet deep, and therefore impossible to be forded. The storm of rain of the night before had so increased the volume of water that the river was, in fact, an impetuous torrent.

A bridge formed of the trunks of trees offered a difficult if not perilous passage to the men, incumbered with baggage, but could not be used at all by the oxen and donkeys, which must therefore be swum over. After a great deal of trouble the oxen did swim to the other bank, but the asses at first refused to follow their example. Partly by persuasion, partly by force, and with a vast sacrifice of time and labour, the negro Barros, aided by two of his mates, succeeded at last in getting them across, the men swimming by their side. The danger of such a proceeding will, however, be appreciated when I tell the reader that the river was full of crocodiles.

More than an hour having been spent in this operation, we pursued our way, marching E.N.E., till we reached the Usserem rivulet, whence I observed, bearing N.N.W., Mount Uba, about which are scattered the hamlets of Caluqueime. We subsequently crossed the river Cacurocác, which runs S.S.W. to the Qué, and half an hour later the river Quissengo, running to the S.E. to flow into the Qué. On the banks of this last-named river we pitched our camp, at four o'clock in the afternoon, near the village of Catonga, where a certain Roque Teixeira has his compound.

Our day's march had been 19 miles, and we were all very much fatigued.

The road was, it is true, in great part on the level, the altitude varying only from 4750 to 4920 feet.

The arborary vegetation was somewhat sparse and meagre, but the herbaceous continued varied and rich.

On the 6th we were again travelling, north-eastwards, and shortly after starting crossed the Qué by a bridge constructed by the natives. This rivulet is 17 feet wide by 3½ deep, and runs S.E. into the Catápi. The latter river, which bears lower down the name of Counge, we reached at 11.30 A.M., and camped upon its left bank. It was found at this spot to be 33 feet wide, and to have a depth of nearly four feet, with a strong current; its course was S.E., and it flows into the Cunene near Lucéque.

I killed this day a large gazelle (*Cervicapra bohor*) the finest of the kind I saw throughout my journey; so large was it that it took four men to carry it to the camp.

As night fell, the dog kept up a constant yelling in the direction of the wood, proving to us that hyenas were wandering round the huts, and when night had regularly set in we had other music in the form of a duet of bass and counter-bass, produced by the roaring of a lion from the undergrowth, and the hoarse grunt of a hippopotamus from the river.

The aspect of the country continued still the same. On the higher ground, stunted woods, with comparatively few tall trees; and below, dells filled with leguminous plants and vast fields or meadows covered with various grasses, through which meandered a peaceful river or rivulet. The soil was still granitic, with rocks of varied aspects cropping from the surface; of mica, however, they contained but little.

We kept on our N.E. course, passing near the

village of Cgassequera, fortified among enormous granite cliffs, and surrounded by gigantic sycamores, producing a singularly picturesque appearance. After passing the Lussóla rivulet, which runs southward to the Catápi, we encamped on the bank of the Nondimba, an affluent, like the former, of the Catápi, but running northwards.

The plateau on which we then stood was a very lofty one, the altitude being found to be 5250 feet.

From this spot we proceeded to Caconda, crossing three rivulets by the way which run N.N.W. to the Catápi and bear the respective names of Chitequi, Gamba, and Upanga. Later on, we fell in with the Catápi itself, flowing W.S.W., and which, it will be remembered, we crossed on the 6th, it then being distinguished by the name of Counge.

At the point we crossed it on this occasion it was 33 feet wide by $3\frac{1}{4}$ deep, and with a feeble current.

Some of the open spaces we had to traverse in the course of the day were covered with stout rushes or canes, springing from a marshy bottom, difficult of passage.

The crossing of the river occupied time, and my companions preceded me to Caconda.

I reached the fortress an hour or two later, and was met at the entrance by the provisional *chefe*, a mulatto, and rich landowner of the district, and sergeant-major of the black forces, who explained that the permanent *chefe* had gone to Benguella, and had left him (the speaker) the bother of receiving us (these were his actual words).

After this most courteous address Snr. Matheus invited me to pass into the fortress. No sooner had I entered the inclosure than I observed talking with my companions a man above the middle height, thin of aspect, with a broad and well-formed head, a somewhat restless eye, wearing a surtout coat and a white cravat, whom Capello introduced under the name of "José d'Anchieta." Yes, there stood before me the first zoological explorer of Africa, a man who had spent eleven years in the districts of Angola, Benguella, and Mossámedes, enriching the cases of the Museum at Lisbon with most valuable specimens. I had subsequently an opportunity of learning his mode of life, which is worthy of a passing notice.

Anchieta was established in the ruins of a church situated at about a couple of hundred yards from the fortress.

The interior of his habitation was in the shape of the letter T, surrounded by broad shelves, on which appeared a confused heap of books, mathematical instruments, photographic apparatus, telescopes, microscopes, retorts, birds of every variety of plumage, flasks of various sizes, earthenware, bread, bottles full of multicoloured liquids, surgical cases, bundles of plants, medical products, cartridge boxes, clothes, and other undistinguishable articles. In one corner was a pile of muskets and rifles of various systems. Alongside the house was an inclosure, wherein I observed some cows and pigs. At the door, sundry negroes and negresses were skinning birds and preparing mammiferi; and among them, seated in an old fauteuil, which showed evidence of long service, and before a huge table, I found José d'Anchieta.

I give up as useless the attempt to describe what was on that table. Of nippers, scalpels, and microscopes there were not a few.

On one side a heap of fragments of birds showed that he was engaged in the study of comparative anatomy. In front of him a flower carefully dissected proved that he had been occupied in determining from the disposition of its petals, the number of its stamens, the shape of its calyx, the arrangement of its seeds and pistil, the names of the family, genus, and species in which it was to be ranked. With his scalpel in hand and his eye fixed on the microscope he is accustomed to pass the hours he can snatch from his labours as a collector, and now it is a flower, now a bird, which forms the object of his studies.

Occasionally his researches are interrupted by the wail of a suffering patient, to whom he devotes the care of a physician or dispenses the medicine necessary for his relief.

Anchieta professes an unbounded respect for Dr. Bocage, the director of the Zoological Museum of Lisbon, and speaks of him with an amount of friendship and esteem rarely met with out of the closest bonds of blood-relationship.

This however is intelligible: Anchieta, who is fully conscious of the services he has rendered to zoologic science, knows that he possesses in Dr. Bocage a man who does him justice, and appreciates those services; a man who completes in Europe the labour the other has begun in Africa; a man, in fine, who knows how much fatigue, how many fevers, how many inconveniences, each one of those specimens has cost its collector.

José d'Anchieta is one of those who merit the respect of all men of science, and more especially of the Portuguese, his compatriots, since, an indefatigable labourer, he has gained honour for his country, while he himself remains respected, though poor, in the midst of the vice and demoralisation by which he is surrounded, and whence he might readily extract profit were he less high-minded and scrupulous.

Merely to speak of him is to utter a eulogium in his favour, as his name at once recalls his labours; and the memory of his works constitutes his praise.

We learned on our arrival that the *chefe* Castro had been superseded and another officer of the army of Africa substituted in his stead. The latter arrived two days after ourselves, and with him Ensign Castro, in charge of the mails from Europe. The avidity with which our letters were devoured may readily be conceived.

I applied at once for carriers, and Snr. Castro offered to accompany me to the residence of José Duarte Bandeira, the principal potentate of Caconda, through whose enormous influence, he said, the thing could be easily managed.

We therefore started for Vicéte on the morning of the 13th, Ivens leaving at the same time for the dwelling of Matheus, with a view to make a survey of the Cunene, at the point of its confluence with the Quando. It was arranged also that I should pay a visit to the same river further southwards.

Capello, who was suffering from a slight attack of fever, was left behind under the care of Anchieta.

My course lay S.S.E., and I speedily crossed the rivers Secula-Biuza, Catápi, and Usongue, flowing W.N.W., and as they are 10 feet wide and more than 3 feet deep the amount of water they pour into the main stream is very considerable.

After trudging some 26 miles in a S.E. direction, I arrived, as night fell, at Vicéte, a fortified compound among rocks, on the summit of a hill which overlooks a vast plain.

I was received by José Duarte Bandeira, who, after a hearty supper, showed me to an excellent bed, of which I stood greatly in need.

The first thing next morning, Ensign Castro broached the subject of carriers, and Bandeira readily engaged to obtain one hundred and twenty, the number we required to help us on to Bihé.

As I expressed my desire to visit the Cunene, it was resolved that we should proceed thither the following day.

We marched 9 miles to the eastward, and fell in with the river at the Porto do Fende.

I had no sooner arrived than I shot a large hippopotamus, which had been imprudent enough to come to the surface of the water to breathe and look about him, within range of my rifle. I spent two days at this place. The river is here some 370 feet wide by 20 feet deep, and has a current of a mile an hour. Its axis at the Porto do Fende is N.W. and S.E. for a space of 2 miles; up stream it runs from N.E. to S.W., and higher up still E. and W., and below it inclines to S.S.W. for 26 miles as far as Luceque. Occasionally its width is as great as 750 feet, and even more.

It abounds in hippopotami and crocodiles.

A mile below the Porto do Fende there are some rapids at the compound called Libata Grande; half a mile further down there are others, known as the Mupas de Canhacuto, and 10 miles lower still are the cataracts of Quiverequeto, the last it can boast of on its upper course, as from that point it is navigable as far as the Humbe.

The right bank, at the spots where I visited it, is mountainous and covered with virgin wood; on the left extends a vast plain from 2½ to 3 miles broad, up to the foot of the mountains, which form a system of slight elevation, running north and south, on whose western slopes are dotted the Fende villages.

At eleven o'clock at night of the 15th there burst over us a terrific storm, with vivid flashes of lightning, and such torrents of rain that we were completely drenched.

We turned our steps once more in the direction of Caconda on the 17th, taking with us a promise to be supplied with carriers in less than a week, and promising in turn to send the following day a keg of *aguardente* to inaugurate the *convocation*. In this part of Africa *aguardente* plays the same part with men as oil does in Europe with machinery. There is no moving without it.

Our host, who had regaled us so well in his own house, most probably forgot that we should have to travel the whole of the day, and that even if we started at daybreak we could not reach Caconda before night. Any way we set off with empty wallets, and by noon we had become desperately hungry.

We came to a halt in an opening in the wood, when I informed my companion, Ensign Castro, that I must positively find something to eat before I took any rest. All I could bag, however, was a quail, which had to serve us both, when cooked in a soldier's pot, for breakfast and dinner. I may frankly avow that I have breakfasted and dined more bountifully than I did on that occasion.

My black fellows, seeing the avidity with which I picked the quail's bones, the dog meanwhile licking his lips and watching my every movement with hungry eyes, made me a present of a root of manioc, which I divided with the ensign.

I reached Caconda at nightfall, and after a capital supper took note that Ivens had not yet returned and that Capello was already convalescent.

Ivens came back on the 19th, and immediately on his return we sent off the keg of *aguardente* to Bandeira, at the same time begging him to use the utmost despatch in getting together the carriers.

On the 23rd, sundry articles that had been ordered arrived from Benguella, and with them six tins of biscuits, a welcome gift, from Antonio Ferreira Marques.

I sent off another messenger to Vicéte, urging

Bandeira to let me have the carriers at once, as we were now waiting for them.

Still the men did not appear, so that I was induced to beg the *chefe* himself to repair to Vicéte, and use both his influence and authority over Bandeira to procure us what we wanted.

The *chefe* went, and shortly after wrote me that sixty-one men were ready and that there would soon be more. He had taken goods with him for payment, but as white calico was the only acceptable currency in those parts, he said that he required fifty pieces more, which we had not got, but which were subsequently advanced by Bandeira.

The day after this communication came another letter from the *chefe* to the effect that the carriers were going to be paid and would come on at once; two days afterwards we had a third letter saying that there were already ninety-four men collected; and finally, on the 5th of February, we received another epistle, informing us that there was not a single carrier ready or ever likely to be!

Imagine our disappointment!

These were early days, so that I had not as yet, out of the depths of my experience, formulated in my mind a principle which later on became with me an article of faith, and my adherence to which, jointly with the King's Rifle, greatly assisted me in smoothing the difficulties of the way and bringing me at length in safety to the end of my journey.

The principle I allude to may be summed up in the following few words.

"In the heart of Africa distrust everybody and everything, until repeated and irrefutable proofs will allow you to bestow your confidence."

The further I went the more fastidious did I become in the matter of these proofs, till I began to class them with the existence of a life-long love, or the solidity of fortune of a merchant with enormous transactions in the hands of distant agents.

On the receipt of the *chefe's* letter, each of us proposed an expedient more foolish and extravagant one than the other, but so great was our annoyance that it is not surprising if our ideas were somewhat confused.

When we had calmed down a little, we determined to hunt about for carriers wheresoever they could be found, and that if the worst came to the worst and none were obtainable, we would start for Bihé without them, and send for our baggage from that place. This last notion appeared to us the most feasible.

The *chefe* came back from Vicéte, but I never could get out of him a reasonable explanation of his and Bandeira's conduct.

It was then resolved that I should start for the Huambo country, to see whether I could get any men from the native chief there, inasmuch as all were agreed, the *chefe* and Anchieta into the bargain, that it was impossible to engage any nearer.

Anchieta informed me that a short time before he had met with great difficulty in sending to Benguella a lot of zoological specimens, things of relatively much easier transport.

The whole of this affair is worthy of careful attention. I learned that not only Bandeira himself, but a certain Mathias, Sergeant Matheus, before alluded to, and others were accustomed to despatch large caravans to distant settlements in the interior, and yet not one of them could obtain a single porter for us!

I began to fancy that there was a fixed determination to throw difficulties in our way, although I did not suspect it so strongly at that time as I had occasion to do later on.

The course of this narrative will show with what

malice aforethought obstacles were raised to my progress, and which, it would almost appear, Providence alone allowed me to overcome.

I will, however, for a time leave this subject in abeyance, and before continuing the account of my adventures, which from this point assume a more striking character, say a few words with respect to Caconda.

The fortified post of Caconda, the deepest in the interior of the district of Benguella, over which at the present time wave the colours of Portugal, forms a square of 328 feet, surrounded by a deep fosse and a parapet, about which, here and there, are distinctly visible the lines of a temporary fortification constructed with some art. An interior stockade forms a second line of defence, and protects a few tumble-down houses, composing the residence of the *chefe*, the barracks, and powder magazine.

Some good pieces of brass ordnance, mounted *en barbette*, and more worn by time than use, expose their green and oxidised muzzles to the approaching wayfarer.

At about 200 yards or so to the south of the fortress are the ruins of a church.

To the north is a group of poor little huts, occupied by the soldiers.

The country round is agreeable, and without being, as is asserted, free from fevers, can boast undoubtedly of having them in a milder form than is observable elsewhere. The population is exceedingly scanty, and has withdrawn itself considerably from the vicinity of the fortress.

The soil is most fertile, and many European plants readily flourish there, and produce abundantly. This I observed to be the case in tiny plots of wheat, potatoes, and other produce.

The Secula-Binza rivulet offers a source of crystalline water purling over a granite bed.

There are but few trees near the fortress, cleared away probably by the necessities of the inhabitants, for there is but little doubt of many having stood here formerly, as they still stand in clumps and woods at some short distance.

Of trade there is but little, and that is carried on very far in the interior.

The same evidence of decline which is visible in Quillengues is still more patent here; yet the importance of Caconda is as great as that of Quillengues, if not even greater; but it presents less security for trading operations, the Benguella road being infested with thieves.

CHAPTER V.

TWENTY DAYS OF PROFOUND ANXIETY.

I leave Caconda—The native chief Quipembe—Quingolo and the chief Caimbo—Forty carriers—Fevers—The Huambo—The native chief Bilombo and his son Capôco—Eighty carriers—Letters and news—All but lost! I move onwards—A knotty question in the Chaca Quimbamba—The rivers Calaé, Canhungamna, and Cunene—A fresh and serious question in the Sambo country—The Cubango—Rains and storms—Serious illness—A terrible adventure—The Bihé at last!

I STARTED from Caconda on the 8th of February, 1878, taking with me six Benguella men, my young negro Pepeca, and Verissimo Gonçalves, to whom I have before alluded; and I was also accompanied by Lieutenant Aguiar, the *chefe* of Caconda, who insisted upon attending me in this expedition, the sole object of which was to make arrangements for carriers. He was probably desirous, in taking this step, to show his willingness to be of service to us, and that he was a stranger to the events that had occurred at Caconda.

I must confess that I never doubted the sincerity of Lieutenant Aguiar, because at that time I had not become so deeply impressed with the truth of the principle I laid down in the foregoing chapter; and even at this day I believe he was as much deceived as myself, notwithstanding his long experience of everything pertaining to these subjugated lands.

After a journey of some 10 miles towards the N.E., during which I crossed a small brook, the Carungolo, near Caconda, and later on the Catapi, which there runs to the S.W., I reached the village of Quipembe, where I was hospitably received by the native chief Quimbundo.

This chief at once sent me a small pig, and as I could not purchase any fowls, he made me a present of one in addition.

In the course of the evening he came to see me at my hut, and after a long conversation took the opportunity of informing me that although his forefathers had always been vassals of the King of Portugal, he himself was not so, inasmuch as the numerous arbitrary acts committed by the various *chefes* against him and his people had snapped asunder all the old engagements; that the White King no longer did him justice; and in choice and even elegant terms he narrated many circumstances upon which he based his accusations against the *chefes*.

The *chefe* was himself present at the interview, and had not a word to say in answer to these accusations against his predecessors, so clearly were they expressed.

My host was a man of no common stamp, and he conversed upon the policy of the Portuguese in Caconda with a degree of judgment difficult to be met with in a provincial negro.

I endeavoured to remove from his mind the bad impression which the *chefes* of Caconda had made upon him, but I fear with very little success. His complaints still further confirmed an opinion I had formed of the unhappy results arising from the miserable stipends bestowed upon the *chefes* of the districts in the interior, a primary cause of the decline of our power and influence in the country.

The native chief of Quipembe is well advanced in years, and moreover suffers from the gout, which makes locomotion with him a matter of some difficulty.

His village is of vast size, well fortified and capitally situated. From the moment of my arrival troops of little negroes and negresses hovered about and regarded me with the utmost surprise, taking to their heels at the slightest movement I made. I endeavoured to overcome the fear which my appearance evidently excited by offering them presents of *grelots* and coral beads; it was with reluctance and trembling that some of the boldest approached near enough to seize the coveted trinkets, and they started away directly these were secured.

My spectacles, and more especially my rug, upon which there figured an enormous lion on a red ground, appeared to be objects of the greatest wonder to them.

On the 9th I quitted the village, travelling N.E., crossed the Utapáira rivulet, and an hour later reached the Cuce, an affluent of the Quando. At this spot the Cuce was found to be 10 feet broad by 6 feet deep, and was difficult of passage, owing to the steepness of its banks and the muddy nature of its bed.

The ground on the right bank presented a gentle slope of no great elevation, whilst on the left it was level to the extent of about five-eighths of a mile in breadth. I passed to the south of the Banja village, magnificently perched on the summit of an eminence, and after crossing three brooks, the Canata and Chitando, which run into the Cuce, and the Atuco, which flows into the Quando, I arrived at the latter river, which I consider one of the great affluents of the Cunene.

The Quando runs southwards, and is here 22 yards wide by 6 to 10 feet in depth.

At the spot where I camped, by the village of Pessenge, the river disappears beneath enormous masses of granite, to see the light once more nearly a mile lower down.

The place presented one of the most charming landscapes I have ever beheld. The banks of the river, which were somewhat elevated, were covered with a luxuriant vegetation, elegant palms springing from the dark green of gigantic thorns. Blackened rocks here and there emerged from the tangled undergrowth, their exposed heads polished through the washing of innumerable storms.

Flocks of small birds twittered and chirped amid the trees; numberless wood-pigeons darted in and out the bushes; and from time to time the grunt of the hippopotamus was heard from the depths of the river.

It was savage beauty in all its power, but marred by one horrible feature in the shape of venomous serpents, with which at almost every step we were brought into proximity.

I killed a few, the bites of which the negroes assured me were mortal.

The appearance of one or two badgers induced me to penetrate the virgin-wood on the left bank of the stream in search of them, when I came unexpectedly upon the ruins of a stone wall, which from its extent might well have encircled some ancient town.

This was the first occasion during my journey of my lying down at night with only the starry sky for a canopy, but I did not sleep the less soundly on that account. I woke at daybreak in time to assist at the destruction of a venomous cobra found wriggling between my bed and that of Lieutenant Aguiar.

At starting we travelled N.E. from the village of Pessenge, and soon reached another, the Canjongo, governed by a petty chief, from whom we obtained a few fowls in barter for some common cloth. We subsequently crossed the river Doroma, an affluent of the Caláe, which runs S.E., and rested for some hours on its left bank, when, resuming our march in a N.N.E. direction, we arrived at five o'clock in the evening at the great village of Quingolo.

The native chief received me hospitably, and at once sent food for my people.

Learning the motive of my journey, he told me that if I had applied to him at the time he would have procured me carriers, but that the *chefes* of Caconda made no account of him, frequently to their own detriment. Even as it was, however, he would supply me with forty men, whom he would despatch to Caconda, and perhaps I might obtain the remainder in the Huambo.

I had here a slight attack of fever. On the 11th, at early morning, the chief called on me and renewed his offer of the forty men, who would, he assured me, leave for Caconda the following day.

I was very desirous of making some purchases of food, but there were no sellers; the chief, Caimbo, on learning this, sent me a fine pig. In return I made him a present of three pieces of striped cloth and a couple of bottles of *aguardente*.

Lieutenant Aguiar resolved to return to Caconda, at which I was very pleased.

At midday the leaders of the carriers who were under orders to march came to receive their pay.

The great village of Quingolo is situated upon a granite mount which overlooks an enormous plain. From between the rocks spring huge sycamores, which lend the place a constant and agreeable freshness. These same rocks, combined with the stockades, make a formidable defence against attack, and the place is rendered stronger by a fosse that runs round it, though it is half choked up. On the very summit of the mount are two gigantic cliffs that form a kind of observatory, from which I saw spread before me one of the most surprising panoramas I have ever beheld.

Of a similar character to the prospect from the lofty cross of Bussaco, if the forest, instead of being confined within the narrow belt of walls, extended from Capes Carvoeiro and Mondego to the sea-coast, scarcely interrupted here and there by verdant glades—the landscape visible from the summit of Quingolo is vaster and more grandiose, its only boundaries being the azure outline of distant mountains, which are too remote to be distinctly visible.

On the 12th, although my fever had increased, I decided upon leaving, and having exchanged the most cordial adieux with the native chief and Lieutenant Aguiar, I resumed my journey at 8.30 A.M., accompanied by three guides furnished me by Caimbo, with whom I parted on the best terms of friendship.

Shortly after starting I passed the Luvubo rivulet, which runs into the Caláe, and at ten o'clock reached the village of the petty chief of Palanca, of whom I solicited shelter, as it was impossible for me to proceed with the fever increasing on me every moment.

Notwithstanding the state of my health, I made some astronomical observations, in order to determine my position, and I mention the circumstance here as it was the first of that series of points which I intended to fix in my passage across Africa.

This hamlet of Palanca was therefore the very first point I laid down on the line which marks my journey from the Atlantic to the Indian Ocean.

Three grammes of quinine which I took during the intermission of the fever produced a rapid improvement, so that I was enabled to go on the following day.

I rode a-straddle on a powerful ox, and kept another in reserve. These animals were well broken in, and made my progress easy; I was able to get a very decent trot out of them, and even occasionally a short gallop.

I started at nearly eight o'clock, and shortly after crossed the river Dôro das Mulheres, the oxen finding it difficult work on account of the muddy bottom.

The heat was intense, and I began to feel extremely ill, so that I called a halt in order to get a little rest.

There were no trees near the place, and I fell asleep upon the baked earth, under a burning sun; my slumbers were of the shortest, and on awaking I had a sensation of freshness, and observed that there was shade. It was caused by the thoughtfulness of my attendants, who were standing around me and sheltering my recumbent body from the ardent rays of a vertical sun. I was touched by such a proof of kindly care.

I again went on, passing a little river, the Dôro dos homens, which unites with the former and subsequently runs into the Caláe, under I know not what name. Two hours later I fell in with the river Gandoassiva, which is nearly 6 yards broad by 3 feet 6 inches deep, on the banks of which I took a rest. It is an affluent of the Caláe, and abounds in small fish, a good many of which we succeeded in catching. My indisposition weighed heavily upon me; extreme weakness was now added to the fever that had reappeared, the former caused through inefficient nourishment, as I had only taken during the last two days a little chicken broth.

I took advantage of the halt to get some good broth made, but it was without salt, as the small provision of that article I had brought from Caconda was completely exhausted.

After a couple of hours' rest, we moved onward, still towards the N.E., and half an hour afterwards crossed the river Cucna, which at that place was $6\frac{1}{2}$ yards broad and nearly 4 feet deep, and was on its way to empty itself into the Caláe.

The Cucna flows between the gentle slopes of lofty hills, but has dug for itself a deep bed, with perpendicular sides, some 7 feet above the water, which made it difficult for the oxen to get across the stream in safety.

The passage cost us two hours' hard labour. A couple of hours later, it being then nightfall, I reached the village of Capôco, the powerful son of the native chief of the Huambo country.

Capôco received me very kindly, gave me his own house for my use, presented me at once with a large pig, and, learning that I was ill, sent me a couple of fowls.

I had some talk with him about carriers, whom he promised to supply.

I made him a present of two pieces of striped cloth and a couple of bottles of *aguardente*. Shortly after, a numerous troop of virgins, recognisable by their bangles of bent wood worn upon the ankles, brought my negroes abundant food in wicker baskets. After taking some lunar altitudes, I lay down to rest in a happy mood, notwithstanding my indisposition, at seeing my excursion so far crowned with success.

On the following day my companions were to join me, and with them I should have not only the society of dear friends and compatriots, but the resources which had now utterly failed me, and of which I stood in such sore need.

I fell asleep, therefore, smiling, nor did any ugly dream disturb my slumbers! And yet I was on the eve of a severe trial—a racking anxiety that was to endure for twenty days.

On the 14th I repaired to the habitation of the father of Capôco, the native chief of the Huambo territory. The village of this chief, whose name is Bilombo, is some 2 miles distant from that of his son, and is situated on the left bank of the river Caláe.

Bilombo, who was expecting me, appeared surrounded by his people, and superbly arrayed in a scarlet tunic, with a chasseur's cap set jauntily upon his head. I handed to him my present, which consisted of three pieces of ordinary striped calico and two bottles of *aguardente*, at which he seemed much gratified. He expressed great surprise at the sight of my Winchester rifle, and requested me to fire it off. His astonishment was very great at beholding me hit a small mark at 230 yards, and it knew no bounds when I broke an egg at sixty paces.

This chief at one time held sway over the entire country of the Huambo, but his power is now considerably reduced. His story may be told in a few brief sentences.

He had married a daughter of the chief of the Bihé, which lady contracted criminal relations with one of his sub-chiefs. The guilty couple managed, however, for some time to conceal their amours from the dusky king. It happened that a misunderstanding occurred between Bilombo and a neighbouring chief, which resulted in a declaration of war. Bilombo assumed the command of his army and departed, leaving the government during his absence in the charge of the very lover of his wife's. The two conspired against the absent monarch, and Capússocússo, for that was the traitor's name, caused himself to be proclaimed king. Bilombo was compelled to yield, and retired to this part of the country, beyond the Calác, where the people still remained faithful to him, and at the period of my journey, as he informed me, he was preparing to take a terrible vengeance on the adultress and her lover.

On my return to Capôco's house, I dismissed the three guides who had accompanied me from Quingolo, and sent letters by them to Capello and Ivens, informing those friends that I was anxiously awaiting them, and bidding them not to part from their loads, as the state of the country was anything but secure.

In the course of the evening I took a stroll along the banks of the Caláe, and was surprised at the quantity of game I fell in with; in fact, I had never seen so much together before, but I killed nothing, being unprepared for such a sight.

The chief Bilombo sent me a present of maize flour and a fine ox, a most valuable gift, and the more treasured as oxen were rare in that part of the country.

The carriers were busy laying in their stock of provisions, with a view to starting next day for Caconda, and I was in the act of writing to my friends, when three porters arrived from the native chief of Quingolo, with letters from them and a basket containing salt and a little bag of rice.

I opened the letters in all haste. Two of them were official and one was private, all signed by Capello and Ivens. They informed me that they had resolved to go on alone, and that, in respect of the forty carriers despatched by me from Quingolo, they sent me forty loads accompanied by the guide Barros, in order that I might convey them to the Bihé.

It was only their imperfect knowledge or utter ignorance of the interior of Africa which could excuse my friends in acting in so strange a manner. I was at that time in a hostile country, and if I had been respected hitherto it was only because the people round me looked upon me and my little band as the vanguard of a considerable troop under the command of the friends in my rear, and the fear of reprisals had, up to that moment, restrained the natural rapacity of the natives. I was in the very district where Silva Porto, the old trader, who was accustomed to traverse with impunity the remotest tracts of country, had frequently to fight his way through hordes of savages eager after plunder.

What would be my fate if it were known that my entire force consisted but of ten men?

I looked my position fairly in the face, and found it replete with difficulties.

Capello and Ivens must have been deceived by some false counsellor, for of a certainty their loyalty would never have allowed them, knowingly, to abandon me in so terrible a position.

Still what was to be done? In three days I might reach Caconda and thence turn back to Benguella. On the other hand I had before me a journey of twenty days to the Bihé, a journey wherein I should have day by day, nay almost hour by hour, to risk both life and property. What should I decide?

The night of the 17th of February was passed in an indescribable state of feverish agitation.

Should I push on? Had I a right to jeopardise the lives of the ten men who were now sleeping so tranquilly around me? Had I a right to risk my own life in so imprudent a venture? Should I return to Benguella?

Who in Europe can estimate the almost insuperable character of the obstacles thus raised to my passage, and which placed me in so dire an uncertainty? Surely none, unless it be a brother explorer who has been as unhappily situated as I then was.

The night was a fearful one; for the fever assisted to worry my brain, and care and anxiety rapidly increased the fever. Daybreak of the 18th found me astir, and there were moments when a phrase forced itself upon my mind and I found myself mechanically giving it utterance.

Audacia fortuna juvat. It was the watchword of the old Romans; it is the law which from time immemorial has dictated the actions of adventurers.

My resolution was taken. I would go on. I had not penetrated into Africa merely to visit the Nano country, however interesting it might be, especially to us Portuguese.

I aroused my men. I put before them in few words the precarious position in which we stood, and my determination to go forward to the Bihé. They one and all assured me of their devotion and their resolve to stand by me to the last.

Of these ten men, three, viz. Verissimo Gonçalves, Augusto, and Camutombo, got back to Lisbon after traversing Africa with me; four followed Capello and Ivens by my orders from the Bihé; one, a negro, Cossusso, went off his head at the Quanza and was entrusted to the care of Silva Porto Domingos Chacahanga, and the two remaining, Manuel and Catraiogrande, fell at my feet, pierced through by the assegais of the Luinas; for, faithful to their promise made on that eventful day, they died in my defence while I was myself defending the national colours.

But at the time the events I am narrating occurred I knew but little of my followers, nor indeed had occasion yet offered to test their valour.

I was still an inmate of Capôco's house, and hitherto he had been lavish of his favours; but Capôco was celebrated far and near as the freebooter of the Nano country, who only a year before had extended his depredations even as far as Quillengues, which he had attacked. What then was likely to be his behaviour when he came to know of my weakness?

Upon him depended the success of my enterprise. Capôco was a man of some four-and-twenty years, of attractive appearance and agreeable manners. Often had Verissimo Gonçalves observed that it seemed impossible he could be the man whose name was a terror to the country round, and whose footsteps, wherever he wandered, were marked by devastation and death. Among his female slaves Verissimo knew several girls who had been stolen from Quillengues during the attack of the previous year. There was one of them with whom I had myself conversed, the daughter of a Quillengues chief, and for whom Capôco demanded a heavy ransom.

Capôco was a man of intelligence, most moderate in both eating and drinking, and although in possession of a large number of female slaves, had a very limited harem.

Amid the barbarism in which he lived and the looseness of his principles, he was not wanting in a certain nobility of feeling. For instance, I observed that the young slave above referred to, a handsome and even elegant girl, wore upon her ankles the wooden bangles which were an infallible sign of virginity; and in my surprise at the circumstance, considering her surroundings, I ventured to ask Capôco how it was he had not made her his own. "I cannot do it," was his reply; "she is my slave by right of war, but so long as her father shows a disposition to ransom her I must respect her, and she shall be respected, for I intend to deliver her up in the same state in which I took her."

One morning Capôco, in talking to me, observed that as Benguella lay over there (pointing to the west), the sun must pass the Huambo before it reached Benguella. I answered that it was quite true, whereupon he wished to know how long it was, after it was born in his country, that it rose upon Lisbon. I tried to make him understand an hour and a half, and explained to him the time that it would take for a man to traverse the distance. This excited his surprise, for, as he told me, he thought our country was very much farther off.

The customs of the people of the Nano and Huambo countries are similar to those prevailing at Quillengues, and they all talk the same language. They work in iron, of which they make their arrows, assegais, and axes, but not their spades, which come from the north.

As I have already incidentally mentioned, the girls, so long as they remain virgins, wear upon both ankles, or upon the left ankle only, certain wooden bangles, and it is considered a great crime if any family should allow its daughters to use such distinctive mark if they have lost their title to wear it.

One custom among these people struck me as very

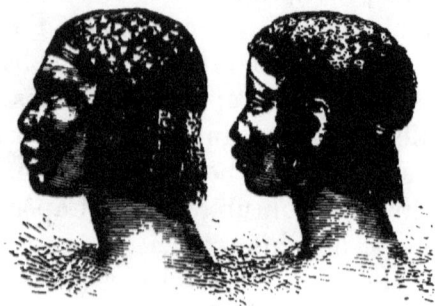

Fig. 4.—Man and Woman of the Huambo.

curious, viz. the existence in every village of a kind of temple for conversation.

This retreat is in the shape of a huge vat, but the ribs which support the thatched roof are placed a good distance apart. In the centre, the hearth is blazing— for the Africans dearly love a fire—and most of the inhabitants of the place, in turn, sit around it on wooden blocks. It is the general meeting-place, more especially when it rains. There one may listen to stirring episodes of war or the chase; love stories are not wanting, nor is there a greater lack than in Europe of tales of wayward lives.

In the country of the Huambo, and on the west coast

thereof, begins the extraordinary luxury of hairdressing, both men and women being remarkable for the style in which they wear their hair; indeed I have seen some heads that it would puzzle the utmost ingenuity of European hairdressers to imitate. Many of these triumphs of the barber's art take two or three days to build up, but on the other hand they last as many months.

The women's hair is profusely adorned with glass beads, which in the trade are known in Benguella under the name of white or red coral, and the article is, of course, greatly sought after in the country. I unfortunately had none of them with me.

Gunpowder, firearms, and table salt are likewise in much request; but these also I lacked, at least in sufficient quantity to part with them; and this only tended the more to make my position embarrassing.

I at length sought out Capôco and told him that my companions had proceeded by the way of Galangue; that only fifty loads would come on, thus reducing the number of men I required to forty, and that I should want them only as far as the Bihé.

On this account we discharged the eighty carriers who by that time were assembled in readiness to start, and who disbanded with many signs of discontent. Capôco promised that I should have the forty I wanted for the Bihé; and on that same day the negro Barros arrived with the forty loads, and another letter from my companions confirming the contents of the first.

From this last epistle I further learned that they had left Caconda for the Bihé, accompanied by the ex-*chefe*, Ensign Castro, and the banished Domingos, who had demonstrated to me the impossibility of obtaining men at Caconda and yet managed to get them himself the very day I left the place.

It was to these last two, in all probability, that I owed

the critical position in which I was now placed, for my companions, but little acquainted with Africa, and not at all with the part of the country in which they were, could not estimate the difficulties which their mode of proceeding had caused me, although the others must have known them full well. I do not of course accuse even them of a crime, but I cannot do less than charge them with great indiscretion.

I do not wish them any ill, for I wish ill to no man, and when, a month after the occurrence of the events I am now recording,—while still suffering from the dangers I had managed to escape, and lying prostrate on my bed, to which I was confined as with iron bands by the sickness following on the twenty days of cruel anxiety of which they were the cause,—I saw them, half famished and entirely resourceless, stagger into the house of Silva Porto, which I then occupied at the Bihé, I freely forgave them all the evil they had done me, and only remembering that one of them was deprived of the rights of citizenship by a sentence which stamped him with infamy, I divided with them the scanty provisions that were left me, and provided them with the means of returning with comparative comfort to Caconda. It may be that I saw in them not merely two white men, two Portuguese all but lost in the remote district of the Bihé, but men who helped to raise me in my own opinion; who, by exposing me during those twenty dreadful days to the numerous dangers which I encountered and overcame, tempered my soul to greater enterprise. Be this as it may, I owed to them the increased trust I felt in Providence and in myself, and in sharing with them the little that I had, I considered that I was paying a debt of gratitude, and rejoiced to think that I had not allowed my sufferings to become a motive for reprisals.

Let me, however, not anticipate facts, but resume the regular course of my narrative.

Capôco called upon me to say that on the following morning I should have the forty men I wanted, but to the Sambo country only, as they refused to go beyond it, owing to the way in which I had dismissed the previous eighty, who had been in readiness to start for Caconda and the Bihé. Besides this, they demanded much higher pay, for whilst I had contracted for ten pieces of cloth from Caconda to the Bihé, the fresh men insisted upon eight pieces from the Huambo to the Sambo. My desire to get away was, however, so great that I thought it prudent to yield.

On the following morning the forty men assembled according to promise; but a new difficulty at once arose. When at Caconda we were bamboozled by Bandeira, Ivens had extracted from all the assorted loads the white calico they contained, because the negroes we expected through Bandeira would not accept payment in any other shape. I had forgotten the circumstance until, on undoing two of the assorted bales, I found that they did not contain a single piece of white stuff. Capôco's people at once declared that they wanted white calico, and nothing but white calico, and that not one load would they lift until they got it.

They refused to have anything to say to the striped, and were actually preparing to leave me, when Capôco himself appeared, and managed, though not without difficulty, to persuade them to take half in striped and the other half in blue.

I saw them off at about ten o'clock, accompanied by Barros, the guide, but they were thoroughly discontented and grumbling. I was myself to follow them in about an hour's time, but had so sudden and violent an attack of fever that I was compelled to delay my journey.

Since the evening before it had been raining in torrents and the night was specially tempestuous.

The fever began to abate about four in the afternoon, and the rain had, by that time, held up. At five precisely I strolled out of the place in the direction of a neighbouring wood, but my steps were uncertain and I had to lean heavily on my staff.

Always liking to be ready for an emergency, I had told my young negro Pepeca, who was in attendance upon me, not to forget to bring one of my rifles. It was fortunate I did so, for we had no sooner entered the wood than an enormous buffalo sprang up within twenty paces of us, and looking at me with fiery eyes, snorted violently.

I took the gun from my attendant's hand, but to my alarm and disgust saw that, instead of a rifle, he had only brought with him a common fowling-piece charged with shot! I felt that it was all over with me, and that death, as inevitable as it was ignominious, was travelling towards me in the shape of yon ferocious beast which was heralding his attack with a low roar.

My thoughts flew towards Heaven, my wife and my daughter. Meanwhile the creature was advancing by leaps, in that irregular way these animals use in making their attacks. At a distance of about eight paces I gave him the first charge of shot. It stopped him for perhaps half a second, and on he came again more madly than before. When I fired the second barrel the muzzle of the gun almost touched the beast's head, and the instant I had done so I leaped nimbly aside. The buffalo turned neither to the right nor left, but continuing his wild career, disappeared in the thicket. Pepeca laughed fit to split his sides and, apparently unconscious of the peril in which we had stood, clapped his hands when he recovered breath, and exclaimed, "The bull has run away! how we must have frightened him!"

I lost no time in returning to Capôco's house after this adventure and passed the night in comparative ease. Before I lay down I wanted to write, and was therefore compelled to improvise a lamp, which I made by sticking some cotton by way of wick into an old sardine box containing pig's lard.

It was on the morning of the 21st February that I took leave of Capôco, and with the fever still upon me wended my way towards the Sambo territory. Before I reached the Caláe I received a note from the guide Barros informing me that during the night the carriers had all fled, leaving their loads in the village of the petty chief Quimbungo, the brother of the chief or native king Bilombo.

I turned back and sought an interview with Capôco, to whom I related what had occurred. He advised me to go on to his uncle's settlement and that he would remedy the mischief. I therefore again proceeded, and shortly after crossed the Caláe, which runs N. and S. to the Cunene, it being at this spot 33 yards wide by 3 feet 2 inches deep, with a violent current.

It flows through vast plains, slightly undulated and clothed with gramineous plants, among which rises, here and there, a solitary dragon-tree. The soil is of animal formation, the whole of the ground being covered, or, more correctly speaking, covering an infinite world of white ants.

A bridge, roughly thrown together and composed of the trunks of trees, unites the two banks of the river. Some 110 yards above the bridge the Caláe receives an important affluent, the Cuçuce, which contributes a volume of water as great as its own. I marched N.E. and at ten o'clock passed close to the village of the petty chief Chacaquimbamba, at the entrance of which there was a large assembly of people. I went by without their saying a word, but had not gone more than 50 yards than I heard a great noise from the direction of the settlement. At the same moment Verissimo came running up to me with the intelligence that one of our own carriers was the innocent cause of the commotion.

I turned back and found the negro Jamba, on whom devolved the duty of carrying my trunk, in a great state of excitement owing to the natives having stolen his gun—a feat which they performed the more readily as, apprehensive of dropping his load, which he knew contained the chronometers and other delicate instruments, he made but a feeble resistance.

Besides the firearm, they had carried off to the village a she-goat and a sheep, a present from Capôco. I gave them to understand that they must restore what had been stolen, but I got nothing but murmurs of a threatening sound in reply.

I made a rapid survey of my position, and did not feel particularly comforted by the reflection that my party consisted of ten men, opposed to upwards of 200.

Urged, however, by a sudden impulse, and putting aside the dictates of prudence and common sense, I determined to test the mettle of those ten men, who were destined to be my comrades in even greater dangers. Moving, therefore, towards the entrance of the village, I cocked my revolver, and ordered them to enter and regain possession of our property. My Benguella negro, Manuel, a young man of whom I had never previously made any account, became, as it were, another being, and cocking his gun led the way at a trot into the village. He was at once followed by Augusto, Verissimo and Catraiogrande, and a moment after by the rest of my troop, leaving me alone to stand the brunt and become perhaps the victim of the fury of the populace. The audacity, however, of our proceeding in all probability saved it from failure, and when Verissimo marched out from the place in triumph with the goat, and Augusto with the sheep, covered by their companions with their guns ready for use, the natives retired to a more convenient distance, and offered no opposition to our movements.

We, however, lost the gun—easier of concealment than the animals, it was hidden securely away ; nor did a second search, which the success attending the first emboldened us to make, bring the missing article to light.

My negroes, heartened by the indecision of the natives, now became loud and warm in their desire for vengeance, and I had to exercise all my authority to prevent them opening fire on the groups that were watching us. I succeeded in calming them at last by a promise of speedy and complete satisfaction at the hands of Capôco, in whom, to tell the truth, I began to feel a certain confidence.

This adventure detained us upwards of an hour, so that it was not till 1.30 P.M. that we crossed the Pöe, an affluent of the Calác, which is five-and-a-half yards wide by nearly four feet deep ; the bottom being soft and muddy rendered it difficult to ford.

At three o'clock we reached the village of the petty chief Quimbungo, brother of the native king of the Huambo, where we found the negro Barros in charge of the abandoned loads. Quimbungo received me very cordially and promised to furnish me with carriers to the Sambo country. On learning also of our adventure of the morning, he begged me not to let my anger fall upon Chacaquimbamba, and he would take care that the stolen gun was restored and full satisfaction given for the insult. About six in the evening Capôco arrived, bringing with him several of the porters who had fled, and the goods which had been given to the others by way of advance of pay. He further told me that on the morrow the gun should be brought back and the chief of the little village be placed at my disposal, that I might inflict upon him such chastisement as I thought proper. And more than this, he assured me that I need no longer fear the flight of any of the carriers, as he himself or his uncle would accompany me as far as the Sambo.

I retired to rest burning with fever, and passed a horrible night.

On the following day a lot more carriers were got together, but still not enough for our purpose.

Capôco started at daybreak for Chacaquimbamba's place, and at mid-day returned with the stolen gun and that chief himself, to whom I graciously extended full pardon for the offence of his people. The delinquent was profuse in his expressions of gratitude and—what was even more satisfactory—presented me with a couple of splendid sheep.

This done, Capôco, the renowned and ferocious chief, the terror of the neighbouring countries, whom I had succeeded in so completely winning to my service that he had heaped me with favours, took his leave, and recommending me warmly to his uncle, quietly returned to his own residence.

As evening fell, a frightful tempest broke over our encampment. Torrents of rain descended amid constant crashes of thunder, and forked lightning darted perpendicularly into the earth all around us. My fever increased amid this war of the elements.

The storm continued with more or less violence throughout the night, but the rain moderated somewhat. Quimbungo, shortly after daylight, informed me that the carriers were ready, but that they demanded payment in advance.

This I positively refused, for besides the experience

recently acquired of the folly of the practice, Capôco had advised me never to pay them beforehand.

The men in turn refused to go, and disbanded. Quimbungo assembled some of his immediate followers and ordered them to accompany me, but the number was very small, so that, even with the addition of those brought me by Capôco, I had still twenty-seven loads without carriers for them, and was compelled to leave them behind under the charge of Barros, Quimbungo promising to send them after me to the Sambo, whither I decided forthwith to bend my steps.

I started at 10 A.M. in an easterly direction, and an hour afterwards crossed the river Canhungamua, 33 yards in breadth and from 13 to 16 feet deep, which running southwards mingles its waters with those of the Cunene.

A bridge of recent construction, formed of the trunks of trees, gave an easy passage to our party, but our carriers on reaching the left bank expressed their determination to go no farther that day. I was compelled to use the utmost energy to make them continue their march until three in the afternoon, at which hour we fixed our encampment in a thick forest of acacia-trees.

The bad weather still pursued us, nor could I throw off the fever which weighed upon me, although it yielded somewhat to the irregular treatment I was enabled to apply.

During the night an awful thunderstorm travelling from south-west to north-east passed over our heads, the vivid flashes of lightning being accompanied by torrents of rain.

Breaking up our camp on the following morning at six, we pursued our journey, reaching the Cunene a couple of hours later. This we crossed by a bridge constructed, like all the bridges in this part of Africa, of unhewn trunks of trees. At this spot the river was found to be 22 yards wide and 6 feet deep, the stream running southwards. The banks are slightly undulated, covered with tall grasses but with little wood. A double row of trees, however, very similar in appearance to the stunted willows of Europe, was traceable by the eye for a considerable distance, in the shape of tortuous lines, between which the river flowed with a rapid current over a bed of fine white sand.

I took a short rest, after making the necessary observations to determine the altitude, and started again at noon, arriving at 2 P.M. at the village of the native chief of the Dumbo in the Sambo territory.

This chief is a vassal of the king of the Sambo, is a man of considerable wealth, and reckons a large number of inhabitants in the villages and hamlets over which he holds sway. He received me very courteously and invited me to take up my quarters within his village, which I accepted.

He promised me carriers for the following day, although, as he said, I had not arrived at a very favourable juncture, as many of his people were absent upon a war excursion. I paid and discharged the Quimbungo carriers and felt confident about resuming my journey on the following day.

A short time before my own arrival, a wealthy chief, by name Cassoma, had reached the Dumbo. He was a friend of my host, whom he had come to visit, travelling for that purpose from his residence on the bank of the Cubango. This Cassoma was far from being sympathetic to me, although he was himself profuse in his expressions of friendship, and even offered to accompany me to the Bihé.

In the evening I sent three bottles of *aguardente* to my host and reminded him not to fail me next morning in the matter of carriers. Contrary to the hospitable customs of the natives in these parts the chief had sent me nothing whatsoever to eat, and as none would sell us flour, we were beginning to get very hungry.

It was about eight o'clock at night that, in a very bad humour and with an empty stomach, I was about to retire to rest, when I heard a knocking at my door, which was immediately followed by the entrance of my host, the chief Cassoma, another by the name of Palanca, a friend and principal counsellor of my host, and five of the wives of the latter.

We conversed awhile about my journey, but Cassoma suddenly broke in with the remark, that they had not come there to talk, and addressing himself pointedly to his friend, he added, " We want *aguardente*, as you know, so tell the white man to give it to us."

My host, encouraged by the impudence of Cassoma, then told me that I must give him and his wives some liquor. To this I replied that I had already given him three bottles, although he had not offered me bit or sup in return; that it was the first time in the course of my travels I had been allowed by a chief, who proffered me hospitality, to go to my bed fasting, and that I should not therefore part with another drop of *aguardente*. Cassoma then took up the cudgels and did all he could to awaken the anger of his brother chief; a warm controversy ensued between us, which lasted for more than an hour, and although I managed to keep my temper, my prudence and patience were tried to their utmost limits.

Patience and prudence, however, alike gave way when my unwelcome visitors declared that, as I would not give them what they wanted by fair means, they intended to help themselves. Pushing the cask towards them with my foot, I seized my revolver, and cocking it, asked who intended to take the first drink.

They hesitated a moment, when Cassoma cried out

to my host, "You are king here, and have a right to the first swill." Dumbo threw off his outer garment, which he delivered to Palanca with the words, "Take care the white man doesn't steal it," and took two steps towards the cask.

I raised my revolver to the height of his head and fired; but Verissimo Gonçalves, who stood by me, knocked up my arm, and the ball went crashing into the wall of the hut.

The three negroes, trembling with fear, retreated to as great a distance from me as the dimensions of the building would allow, and the five women set up a horrible chorus of screams.

I then for the first time became conscious of the sound of other human voices mixed with that laughter so peculiar to the blacks, and looking towards the door I discovered my faithful followers Augusto and Manuel who, on hearing the discussion, had softly approached, with the rest of my men in the rear, and now, armed with their guns, were keeping guard at the entrance, and heartily enjoying the scene.

Verissimo then, in a confidential tone, informed my host and his companions that they had better retire and not say a word to arouse my anger, for that if I should put myself in a rage again he would not answer for the consequences or be able perhaps to save their lives, as he had done awhile ago.

They lost no time in taking his advice and filed off, one behind the other, in the utmost silence.

But for Verissimo's knocking up my arm in the way he did I should have killed the chief, and in the position in which we then stood we should in all probability have been massacred to a man. In saving my host's life, he had therefore saved the lives of us all.

The excitement occasioned by this last adventure so increased the fever within me, that when the place was cleared of my visitors I dropped in a state of utter exhaustion upon the skins which, spread in a corner of the hut, served me by way of bed.

My faithful blacks stretched themselves across the door and told me to sleep in peace, as they would watch over my safety.

On three different occasions, therefore, within four days had my life been in jeopardy. First, in my encounter with the buffalo in the Huambo; secondly, in the forcible entrance into Chacaquimbamba's village; and thirdly, in the adventure of that evening.

After a short and broken sleep I awoke to the sounds of a tempest that was raging violently outside.

As I lay, I turned over in my mind the events of the few hours before, and did not derive much comfort or tranquillity from their contemplation. What would the morning bring forth? There was I, with my ten men, within a fortified village whence it was not easy to escape, and even were the passage clear, where was I to obtain carriers now that I was, so to speak, at daggers drawn with their chief?

My readers may form some slight idea of the anxiety with which I watched for the first gleam of daylight.

When the dawn at last appeared I took it as a good omen that the fever had somewhat abated. I rose, made all preparations for departure, and then took the bold course of summoning the chief, who was not long in making his appearance.

I told him that I was about to continue my journey, and should leave my property under his care, until such time as I could send for it. In a very subdued manner he begged me not to do that, as he would furnish me with carriers; he made a thousand apologies for the occurrence of the evening before, the whole blame of which he threw upon Cassoma whom, as he averred, he had turned out of his house. This, however, was not true, as I caught a glimpse of the fellow a little later on.

At ten o'clock the requisite carriers appeared. But I saw at a glance that they did not all deserve that name, for amid the group were half-a-dozen girls with bangles about their ankles; so that, in his hurry to get rid of me, he had not waited to draw men from the surrounding hamlets, but put all he had at my disposal and made up the desired number by these female slaves.

Fig. 5.—Woman of the Sambo.

I, however, thanked him warmly and expressed my satisfaction at such a proof of courtesy, adding that I had not got with me a present worthy of his acceptance, but that I should be happy to offer him a handsome gun if he would send a man with me, in whom he placed confidence, to receive it at the Bihé; hinting, at the same time, that I should be pleased if he selected for such office his confidant, the chief Palanca. My delight was extreme (though I took care to conceal it) at his yielding to my request and appointing Palanca to accompany me. By so doing, this Dumbo princelet delivered into my hands a precious hostage, who would be responsible not only for my own safety but for that of the loads I had entrusted two days previously to the care of Barros, whom I informed of the circumstances by a letter which I left for him at the Dumbo.

I quitted the village, which had so narrowly escaped becoming a scene of successful treachery and bloodshed, at 11 A.M., marching at the head of my strangely assorted crew, consisting of my ten Benguella braves, ten very doubtful characters of the Sambo country, and six virgin slaves of the native chief of the Dumbo.

The rain was falling in torrents; but heedless of

this inconvenience I trudged steadily on, anxious, as may well be supposed, to put as many miles as possible between myself and that inhospitable township.

Four hours later, having travelled N.E., I pitched my camp near the village of Burundoa, completely soaked through and shivering with cold and fever.

I declined the hospitality offered me by the chief of the locality, for not only had I been vividly impressed with the experience of the evening before, but I began

Fig. 6.—My Encampment between the Sambo and the Bihé.

to see the wisdom of the counsel given me by Stanley, namely, never in Africa, if it could possibly be avoided, to pass the night under native roofs.

Several girls made their appearance at my camp, offering for sale Indian corn, both whole and in flour, and some magnificent potatoes, in no way inferior to those of Europe.

Rain still continued falling—less heavily, but most persistently—and I really began to feel very ill.

In the vicinity of my camp there was a little brook, whose waters helped to swell a rivulet, an affluent of the Cubango, into which it flowed somewhat farther to the westward.

During the night the rain kept falling, and increased in violence between four and five in the morning, at which latter hour it held up. There is great abundance of excellent tobacco in this country, where a good deal was sold me at a very cheap rate. Few of the blacks, however, in those parts seem to smoke, but all use tobacco in the shape of snuff. This they prepare in a very primitive way, by roasting the leaf before a slow fire and then pounding it in the very tube or box, out of which they take it, by means of a little wooden pestle fastened to the box by a fine strap.

I started at 7.40 A.M. in a N.E. direction, traversing a highly-cultivated and thickly-peopled region.

At 8.30 we passed close by the large hamlet of Vaneno, and at ten made a short halt close to the village of Moenacuchimba. We resumed our march half an hour afterwards, still pursuing a N.E. course; at eleven were abreast of the hamlet of Chacapombo, a very populous place, and at 11.30 had another rest near Quiaia, the most important of all these inhabited places.

The chief of this latter village turned out to salute me and made me a present of a large pig. I returned him its value in striped cotton stuff, at which he was very pleased, and subsequently sent a lot of pumpkins for the use of my people.

We pursued our journey in the same direction, and two hours later pitched our tents in a wood near the hamlet of the Gongo. The latter part of this day's march was very tedious owing to the heavy showers of rain; and a S.W. wind that was blowing was searching and cold.

In the evening an envoy arrived at my camp from the native chief of the Sambo, whose township was described to me as being situated at a distance of some nine or ten miles in a N.W. direction. The object of his message was to get something out of me in the way of a present, and to inform me that if I would pass by the chief's place he would give me an ox in return. I thanked him for the kind intention, and promised to let him have a trifle on the following day, for I was apprehensive, if I sent him off empty-handed, he would induce my carriers to abandon me; a matter that it would have been very easy to do, as they had already shown a disposition to mutiny which it had required all Verissimo's eloquence to overcome.

A chief of the name of Capuço, who held sway over the neighbouring hamlet, paid me the compliment of sending me by three of his wives (all very ugly women) a present in the shape of a fowl and three pumpkins. In return I sent him about three yards of striped cloth and gave a few beads to the women. At nightfall we had other female visitors, offering flour, maize and manioc for sale. All these women indulged in the most extravagant head-dresses, the hair being interlaced with white coral and made to shine with a lavish expenditure of castor-oil, which seemed to be a favourite article of the toilet.

The men furnished me by the chief of the Dumbo were the most insubordinate rascals I ever came across; they were always either quarrelling with one another or with the Benguella porters, so that the only quiet spot in the camp, at night-time, was that occupied by the six negresses, my gentle virgin carriers.

A very rough night it was—rain and wind contending for the mastery. At daybreak the chief Capuço came to thank me for the cloth I had given him, and as if to make up for the insignificance of his former gift, had brought with him a handsome pig and a good fat hen.

The envoy of the great chief came shortly after to receive the present I had promised him; and as I considered it was only an exchange for an *intention* to give me an ox, if I went ten miles out of my way, I did not think it worth while to make it a costly one.

At 8 A.M. we were on our way, and at 9 passed close to the hamlets of Chacaonha, inhabited by the first of the Ganguella race in West Africa.

The Bomba rivulet was shortly after forded and we continued along its left bank for about a mile and a quarter, when the carriers suddenly laid down their loads, saying they would not move another step, and demanded payment that they might return to their homes. We were then about a mile or so from the Cubango, and being very desirous of crossing that river, I tried to persuade them to go at least that short distance farther, and promised that, so soon as I was on the other side, I would pay them what was due and dismiss them.

My persuasions, however, had no effect. They gave me to understand that the reason of their refusal was the fear of my vengeance—that I had been grossly insulted in the village of their chief at the Dumbo— and they were convinced that I should not spare them if I once got them on the other bank of the river and consequently out of their own territory.

I tried to reason them out of such an absurdity, but it was labour in vain. I then refused to pay them at all if they did not carry the loads to the other side of the river. To this they replied that they would rather go without their pay than follow me, and they at once called the six girls and bade them come away with them.

I was at my wits' end. Within a stone's throw, as it were, was the hamlet of that fellow Cassoma, and I thought I perceived in this business a craftily devised plan to betray me into his hands, he having gone on before to make his preparations.

Any loads abandoned in such a place were as good as lost beyond redemption, and with this conviction on my mind my readers may imagine with what feelings I contemplated the departure of the carriers.

I turned my eyes, in perplexity, towards my goods, and a sudden revulsion of feeling came over me. Seated on one of the packages that were spread upon the ground was a tall, thin figure of a man, with a face as immovable as if cut out of stone, and with a long gun lying across his knees. It was the petty chief Palanca, who had accompanied me from the Dumbo, and whose existence I had almost forgotten. Now or never was the time I could make him useful. Making a spring upon him, I disarmed and threw him to the ground. Calling to my men, I ordered them to bind him hand and foot, and in a loud voice commanded Augusto and Manuel to hang him up to the projecting branch of an acacia which conveniently presented itself for the purpose.

Seeing by the rope put about his neck that the order was being most undoubtedly carried out, the fellow exclaimed: "Don't kill me, don't kill me; the carriers shall go across the Cubango!" at the same time he gave vent to a loud halloa which brought back the men, who were already at some little distance.

When they were reassembled he gave the word for them to take up their loads and follow him, a command which they obeyed without hesitation.

I then ordered that his feet should be unbound, and threatened him with a bullet through his head at the slightest mutiny of the carriers. Half an hour afterwards we passed the Cubango by a well-constructed bridge, and camped on the left bank near the hamlets of Chindonga.

I found between the river and my camp some iron mines whence the natives extract abundant ore.

At length I stood in the Moma country, and free of the territories of the Nano, Huambo and Sambo, of which I shall retain a life-long memory.

The Cubango there runs to S.S.E. and is 38 yards wide, by 6 to 13 feet deep. I made some observations to determine the position and altitude, but was forced

Fig. 7.—Cassanha Bridge over the River Cubango.

to take speedy refuge in my hut, as a squall from N.N.E. discharged upon me a copious amount of rain.

I paid and discharged the Sambo carriers, giving them a yard of striped cloth each, which was the recompense agreed on.

I then called the six girls and told them I should give them nothing, as women were bound to work, and deserved no pay. They hung their heads in a very downcast fashion, but made no remark at my decision, so degraded is the position of women in this part of the world.

Just as they were about to start, and had turned their heads towards the Sambo, I ordered them to come back, when I made each of them a present of a couple of yards of the most brilliant chintz I possessed, and some strings of different beads.

It is impossible to describe the delight of these poor creatures at receiving so splendid a gift. The men

Fig. 8.—THE SECULO WHO GAVE ME A PIG.

looked on in envy, and I improved the occasion by pointing out to them that, if they had not mutinied on the other side of the Cubango, I would have given them the same guerdon.

This was my revenge, and I hope the lesson was not lost upon the fellows.

In the course of the evening a petty chief from Chindonga came to visit me, bringing with him a pig as a present.

He promised me carriers for the following morning, at the rate of half a yard of striped cloth per day, telling me, however, that they would only go as far as the Caquingue country, where I should readily obtain men for the Bihé.

My fever had yielded to the tremendous doses of quinine I had taken; but, completely wetted through for three whole days, I began to feel the first symptoms of that rheumatism which threatened more than once to bring my journey to a sudden close.

The night was tempestuous, and the following day continued very wet.

The chief was as good as his word, and put in an appearance early next morning with the carriers; but I had resolved to give myself some hours' rest, and therefore dismissed them till the following day. I learned from the chief that my companions had passed through his place on the previous eve, coming from the south.

The chief, Palanca, from the Sambo was carefully watched, but was otherwise free. The day before I had despatched a message to my former host of the Dumbo, informing him that the head of his friend should answer for the loads that had been left behind in the care of Barros, a resolution which Palanca found most just and natural, as it was the law of the country.

It is not improbable that this, and other proceedings of mine, which will be found most frankly avowed in the course of this narrative, may be censured by some of my readers; but I would beg my censors to ponder for a moment upon my position, accompanied as I was by a mere handful of men, in a country where everything was hostile, climate and inhabitants included. If I do not profess the principle that the end justifies the means, neither do I lay claim to that virtue which would present the other cheek when the first has been smitten.

Far from the restraints of the civilised world—outside its two circles of iron—the penal code and social conventionalities, which, close and rigid as they are, still leave sufficient room for crime and infamy, the African explorer, hemmed in by savage races whose rules of conduct differ essentially from his own, having the Almighty as sole witness of his acts, and his conscience as sole censor of his proceedings, requires a more than ordinary strength to preserve his honesty of purpose and moral dignity amid scenes and circumstances where his passions might so easily lead him astray. For myself, I candidly confess that the ovations which have been showered on me by the civilised world, for having happily overcome the material obstacles of my journey, might have been perhaps more justly bestowed upon me for my victories over my own self, if the terrible internal struggles I had to undergo had only been as patent to the eye.

To conquer his own unruly passions, to overcome the material and moral habits he has formed during his civilised life, are the two great labours of the explorer. He who can do this successfully will attain his end and fulfil his mission.

At the outset of my journey I must confess I had some apprehensions on this score, and as time went on I discovered that my fears were not unfounded. I had to wrestle severely with my own spirit, but though exhausted with the struggle, I managed to come out victorious. By dint of indomitable will, I succeeded in establishing an empire over myself, and though lacking time to produce a written code of conduct, I formulated one in my mind by which I guided my proceedings. My principles were those of natural right; my law, brief but excellent, was summed up in the ten precepts of the Decalogue.

Let it not be for an instant imagined that I put

forward any claim to canonisation, or that I pretend to have rigorously followed the precepts laid down in the twentieth chapter of the sublime Book of Exodus, certainly the most beautiful of the Pentateuch; but I did my best not to depart too widely from them, and in so doing I did well.

If this digression do not greatly help on the narrative, it may at least be useful in awakening some chord in the heart of future explorers, and to them it is in all heartiness addressed.

Fig. 9.—GANGUELLA WOMEN ON THE BANKS OF THE CUBANGO.

To resume.

During the day a great many negroes came about us offering for sale various articles of food, of the usual kind, but there was one comestible which was singular enough to deserve a passing notice.

A large basket displayed a quantity of caterpillars, very similar to the *Acherontia Atropos*, and of the same size. This gigantic lepidopter, when young, feeds upon the grasses and is then easily caught. The Ganguellas devour it ravenously, but my own men refused to touch it.

On the following morning, at the first appearance of daylight, a good many more carriers presented themselves; but, as I had already my number, I was compelled to dismiss them.

I left about ten o'clock, by which hour the rain had fortunately held up. Just as I was starting I had the ill-luck to break my spectacles, which I had worn ever since I left Lisbon. Our course was N E., and after five hours' tramp we pitched our tents on the left bank of the river Cutato dos Ganguellas, the stream being passed by stepping-stones a little above a small cataract.

On the road we forded a petty brook called Chimbuicoque, an affluent of the Cutato.

At that point the river runs eastward, bending subsequently to the north and then east by south. This gigantic S is a series of rapids, where the river rushes with a tremendous roar over the granite rocks which form its bed.

At the site of the stepping-stones, or natural stone bridges, it measures 88 yards across, and about 30 yards both higher up and lower down, with a depth of 13 to 16 feet. It flows into the Cubango, so say the natives, at fifteen days' march to the south of this point.

The right bank is covered with the plantations of the inhabitants of Moma, which occupy a space that I calculated roughly at upwards of two thousand acres of land. They are the largest I have ever seen in Africa. The crops produced by these people consist mainly of maize, beans and potatoes, but maize fields are those which chiefly meet the eye. Before reaching the plantations, I crossed a forest of enormous acacias of surprising beauty. The aspect of the banks of the Cutato is very singular. Where the granite of the river-bed terminates, a soil commences of termitic formation, the ground undulating in thousands of little hills, some cultivated, others covered with sylvan vegetation; and as they are all connected, the aspect is that of a system of miniature mountain chains which perfectly enchant the beholder. I fixed the position of the large village or township of Monu at the distance of two miles bearing W.S.W., and after determining the altitude of the river there, I sought my tent, wet through from the incessant rain, and with another attack of fever upon me.

Threats of rheumatism continued. During the night the rain came down in torrents, and, as was

Fig. 11.—Ant-hill 13 feet high on the Banks of the River Cutato dos Ganguellas, covered with Vegetation.

constantly the case now, I went to sleep in the wet, for at this period of the year the grasses, with which I covered my roughly constructed hut, were never more than some 20 inches long, and with such short stuff it is difficult, if not impossible, to keep the water out.

It was not till noon on the following day that the rain ceased; and though my pulse was going at the rate of 144 per minute, from fever, I resumed my journey at 2 p.m.

I tramped along on foot, as I found it impossible to keep my seat on the ox; but after an hour's march my legs refused to carry me farther. We therefore

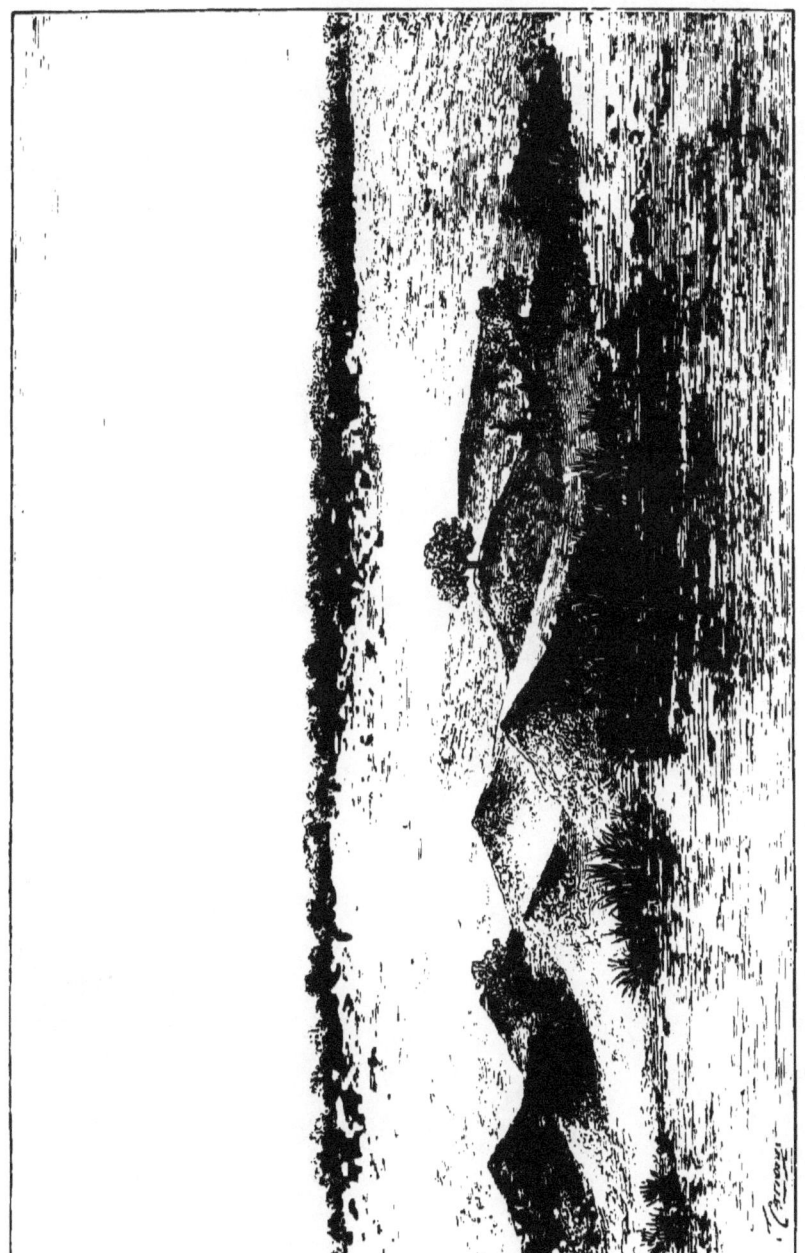

Fig. 10.—ANT-HILLS ON THE BANKS OF THE RIVER CUTATO OF THE GANGUELLAS.

camped; and I met with the utmost attention and care not only from my own negroes, but even from the Ganguella carriers.

The spot where we rested was near the hamlet of a tribe called Lamupas, from their residing near the cataracts of the river, which in the language of the country are styled Mupas.

It is very thickly peopled and extensively cultivated, as the inhabitants are greatly devoted to agriculture.

On my road I fell in with several graves of the native chiefs, which are covered with clay, similar in shape to many in Europe. These graves are protected from the rain by a species of open shed with thatched roof, and are always shaded by a large tree.

Upon most of them I saw earthen vases and platters, placed there by the relatives of the deceased, as we are accustomed to deposit garlands and immortelles upon the tombs of our own loved ones.

Towards night the rain moderated, and on the following morning it was misty but warm. The fever had considerably diminished, but my rheumatic pains began to worry me excessively. Still I went on, and half an hour after having left the camp I passed near the large village of Cassequera.

After crossing a little brook which ran on the other side of the village, I came upon some enormous clearances covered with grasses, which excited my attention on account of their huge size and mature growth at a period of the year when the plants of this family are only just beginning to develop.

My young negro Pepeca had so violent and sudden an attack of fever that he sank down powerless. I called a halt, and sent off a messenger to the village of Cassequera to hire a man for the purpose of carrying the poor fellow on his shoulders. At noon I passed near the residence of the captain of the Quingue, the first village in the Caquingue country. I took up my quarters in the house of João Albino, a half-caste of Benguella, the son of the old Portuguese trader Luiz Albino, who was killed by a buffalo in the wilds of the Zambesi.

João Albino resides in the compound of Camenha, son of the captain of the Quingue.

Camenha himself was absent, having gone to take the command of the forces of the native king of Caquingue, in a war then waging with certain chiefs of the Cubango.

Fig. 12.—Tomb of a Native Chief.

The weather improved and my fever entirely left me, but I had not got rid of my rheumatism, which gave lively evidence of its presence.

It is worthy of note that night came on without rain, and was followed by a cloudless morning.

I paid a visit to the old captain of the Quingue, taking with me, by way of offering, a piece of linen cloth. He made me a present of an ox, which I ordered at once to be slaughtered, as we had eaten no other flesh than that of swine for a long time past. The captain was very old and infirm. He conversed with me at great length about my journey and its motives, and could not comprehend what I intended to do.

When I was about to leave him he said, " I know now who you are; you are a chief of the white king, and he has sent you to visit these parts, and study the roads; for the white king knows that many things are done here that are not good, and he wants to put a stop to them. I pray you, when he does so, not to forgot that I gave you an ox, and treated you as my brother. I have not long to live, but then you can remember my sons, and will do them, I hope, no injury."

I was touched by the old man's words. His chiefs accompanied me respectfully to the village of the son, where I was lodging, and there were few of them who failed, during the day, to bring me over some little present, such as a hen or two, some eggs and sugar-cane. I saw a small plantation of the latter within the captain's enclosure, of even a more flourishing character than that visible on the sea-shore, where this plant nevertheless assumes colossal proportions.

I mention this circumstance, because I was under the impression until then that, at so considerable an altitude, nearly 5580 feet, the cane would not grow.

On my return to the village, I found Francisco Gonçalves, known as Carique, the half-brother of my follower Verissimo, who, learning of my arrival, had come to pay me a visit.

This Carique was, like Verissimo, the son of the trader Guilherme, but by a different mother, and on the mother's side he was heir to the throne of Caquingue.

He lives with the native king, his uncle, and is married to a daughter of the future sovereign of the Bihé.

He was educated at Benguella, and has some sort of culture and a good deal of intelligence. He brought several negroes with him, slaves of his father, whom he placed at my disposal to accompany me in my journey eastward from the Bihé country.

Thus, before I had even reached that desired goal, I had several carriers in readiness.

Carique, Albino, the captain's son and others who trade with the interior, start from that point for the Mucusso and Sulatebelle, descending by the Cubango to the Ngami, always on the right bank; and they do business also in the Cuanhama, a country to the east of the Humbe, on the left bank of the Cunene.

Their staple article of trade is slaves, exchanged, on the road, for oxen; and these again, with bale goods, are bartered for wax and ivory.

I resolved to remain there a day, not only to get a long rest and dry my wetted things, but also to procure some information about the country, whose customs differ considerably from those of the tribes I had hitherto met with. In the evening Carique and João Albino kept me company, and furnished me with lengthy data concerning the territory and its people, the most noteworthy of which I here transcribe from my diary.

The Caquingue country is bounded on the north by the Bihé, on the west by the Moma territory, and on the east and south by confederate tribes of the Ganguella race. This latter race occupies in this part of Africa a vast tract of land, and is divided into four large groups, which are susceptible again of further subdivision. Their language and customs are the same throughout, but there is a difference in their political organisation. In the Caquingue country the Ganguellas assume the name of Gonzellos, form a separate kingdom, and admit but one sole head.

In their other divisions they form confederations, which are very common in Africa, each large village or township being governed by an independent chief. Those who live to the S.E. of Caquingue style themselves Nhembas; those to the south, Massacas; and they who dwell to the east of the Bihé, Bundas. Of the last mentioned I shall have occasion to speak at some length later on. The Gonzellos, the Ganguellas of Caquingue, are cultivators of the soil and traders,

Fig. 13.—Caquingue Blacksmiths.

and, of all the peoples of South Central Africa, are those which approximate most to the Bihenos in the way of commercial exploration.

When at home they work a good deal in iron, and this branch of trade establishes between them and other tribes very active commercial relations.

They have not the slightest idea of any religion whatsoever, and though thorough believers in sorcery, they never give a thought to the existence of a Supreme Being, by whom all things are ordered.

During the coldest months, that is to say June and July, the Gonzellos miners leave their homes, and take up their abode in extensive encampments near the iron-mines, which are abundant in the country.

In order to extract the ore, they dig circular holes or shafts of about 10 to 13 feet in diameter, but not more than 6 or 7 feet deep; this arises most probably from their want of means to raise the ore to a greater elevation. I examined several of these shafts in the neighbourhood of the Cubango, and found them all of a similar character.

As soon as they have extracted sufficient ore for the work of the year, they begin separating the iron. This is done in holes of no great depth, the ore being mixed with charcoal, and the temperature being raised by means of primitive bellows, consisting of two wooden cylinders about a foot in diameter, hollowed out to a depth of 4 inches and covered with two tanned goat-skins, to which are fixed two handles, 20 inches long and half an inch thick. By a rapid movement of these handles, a current of air is produced which plays upon the charcoal through two hollow wooden tubes attached to the cylinders, and furnished with clay muzzles.

Fig. 14.—1. BELLOWS. 2. CLAY MUZZLE. 3. ANVIL. 4. HAMMER.

By incessant labour, kept up night and day, the whole of the metal becomes transformed, by ordinary processes, into spades, axes, war-hatchets, arrow-heads, assegais, nails, knives, and bullets for fire-arms, and even occasionally fire-arms themselves, the iron being tempered with ox-grease and salt. I have seen a good many of these guns carry as well as the best pieces made of cast steel.

TWENTY DAYS OF PROFOUND ANXIETY. 129

During the whole of the time that these labours last no woman, under any pretext, is allowed to go near the

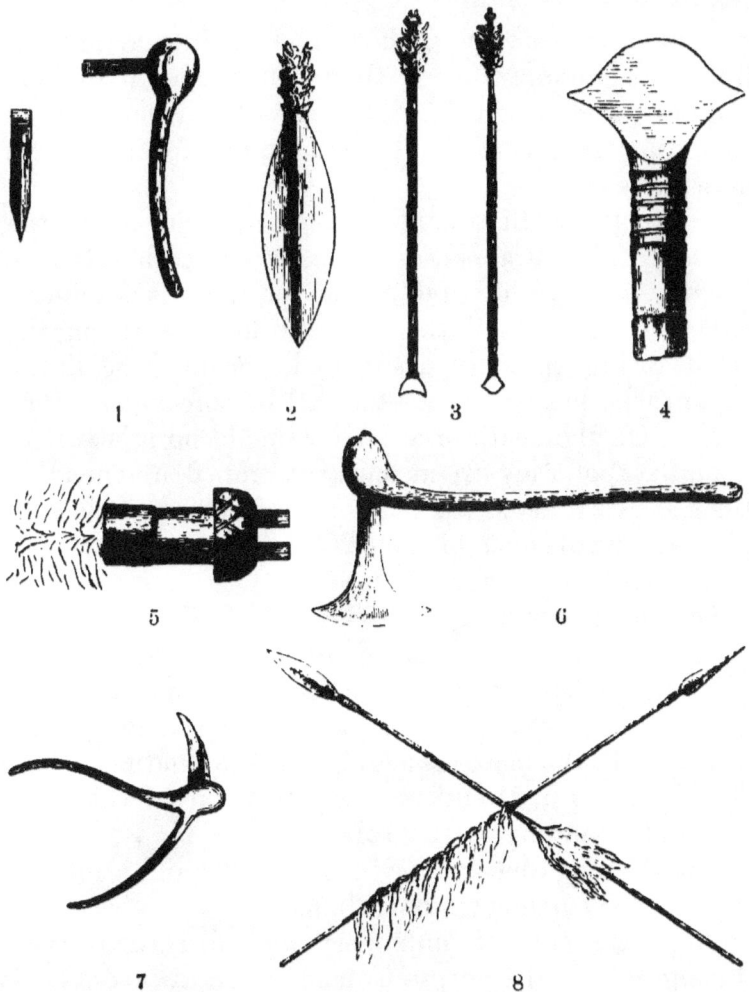

Fig. 15. — Articles manufactured by the Natives between the Coast and the Bihé. 1. Working Axe. 2. Arrow-head for War. 3. Arrows. 4. Arrow-head for Hunting. 5. Butt-end of Arrows. 6. Battle-axe. 7. Hoe. 8. Assegais.

miners' camp, for fear, as they say, of the utter ruin of the metal. My own opinion is, that the object of the prohibition is to prevent the men being distracted in their work, which, as I have stated, is kept up night and day.

So soon as the metal is converted into articles of trade, the miners return to their homes laden with their manufactures, which they subsequently dispose of by sale, after reserving what they require for their own necessities.

It is curious that none of these people admit the existence of natural causes of disease or death. If any among them should fall ill or die, the cause is attributed either to the souls of the other world (one among the spirits being specially designated), or to some living person who has compassed the evil by sorcery or witchcraft. On the death of a native, should no relatives be upon the spot, they are at once summoned, and pending their arrival, the corpse is suspended from a stout pole, planted at a distance of some 200 or 300 yards from the entrance of the village.

On the assembling of the relatives, divination is at once resorted to in order to learn the cause of death. For this purpose the corpse is fastened to a long stake; a man seizes each end of it, and the body is thus conveyed to the place set apart for the divination, where the diviner is in attendance, together with a concourse of people standing in two rows.

The diviner then taking in his right hand a piece of white coral commences operations.

After no end of mummery and discordant cries, during which the corpse is made to sway about—the people all the while believing it does so without human intervention—the diviner declares that it was the soul of such a person, male or female, whom he mentions, that occasioned the death; or he avers that it was this or that *living* person who slew the defunct by sorcery.

In the former case, a grave being dug in the neighbouring wood, no spot in particular being selected for the purpose, the body is interred without more ado, and stones, wood and earth are heaped over it. But in the latter case, the person designated by the diviner as the sorcerer is seized, and must either pay to the nearest of kin the value of the life he is deemed to have taken, or forfeit his head; an account of the event being subsequently given to the ruling chief, together with a female goat as a fee for listening to the case.

An accused person has fortunately the right to deny his supposed crime, and to furnish a defence. He applies for such purpose to a medicine-man (by way of advocate), who, in presence of the people, proceeds to prepare his proofs, in the shape of an ordeal, to establish either the guilt or innocence of the accused. For instance, in sight of the latter's kinsfolk and of the general public, he composes a poisonous draught, to be taken both by the accused and the nearest relative of the dead man. This draught produces a species of temporary madness, and he who suffers most from its effects is deemed the more guilty, and has sentence of death passed upon him.*

If this sentence fall upon the accused, he either pays the life of the deceased, or forfeits his own; if it fall upon the other man, he has to indemnify the accused for the accusation made by giving him at once a pig, to pay for the trouble in seeking a medicine-man, and subsequently, whatever else the accused may claim, namely, a couple of oxen, two slaves, a bale of goods, &c., &c.

In this place, I cannot do better than point out a vast difference which exists between three important personages among the people of South Central Africa, and who

* This is very similar to the practice in use among the Maraves, the Ordeal of the Muave (Gamito, Muata Cazembe).

are not unfrequently confounded with each other. These are the medicine-man, the diviner, and the sorcerer. At the first glance they do certainly appear to have points of contact, but in reality they have none.

The medicine-man is defined by the name bestowed upon him. He prepares medicaments. He has some knowledge of medicinal herbs and roots which he invariably employs empirically, and makes great use of the cupping glass; but as regards science, he has little or none. The medicine-man never makes a diagnosis of any disease, but deals freely in prognostics. His doses of medicinal plants are always empirical, and the most absurd and useless components enter into his pharmacopœia. It is true that amongst ourselves the use of antidotes does not go very far. The medicine-man, who is at the same time a compounder of drugs, employs during their preparation a certain number of ceremonies and words without which they would lose their virtue. He makes a great secret of the plants and simples he uses, and puts on a very sapient air when questioned upon the subject. The medicine-man is a person of great importance, and many solemn acts require his presence. He decides many great questions, his opinion prevailing over that of the diviner (Ditangja), and he never pronounces it without a preliminary flourish, in the shape of remedies and ceremonies, performed now with plants, now with the blood of human creatures, or beasts, and on which are bestowed the name of *medicinal rites*.

The diviner, on the other hand, deals in divination and nothing else. In the case of any one falling sick, the diviner is first called in to divine whether the attack is due to spirits of another world, or to sorcery, and it is after his work is done that the medicine-man is applied to.

These two personages always perfectly understand each other.

The diviner is not consulted solely in cases of disease or death, he is appealed to in all conceivable matters of moment, and nothing is done without his being first called in.

In questions of consultation, he takes up his stand in the centre of a circle formed by the people, who must be seated. He brings with him a calabash and a basket. The calabash contains large glass beads and dried maize; the basket is full of the queerest odds and ends, such as human bones, dried vegetables, stones, bits of stick, the stones of fruit, birds' and fishes' bones, &c.

He begins by shaking the calabash about in the most frantic way, and during the rattle consequent on the operation he invokes the *malignant spirits;* the basket is then shaken up, and in the articles that appear uppermost be reads what his hearers are desirous of learning of the past, present, or future. I found this ceremony to be in use from the time of my leaving the coast, but in no instance carried out so completely as here.

I spoke of *malignant spirits,* and must explain that in this part of the world such spirits seem to be on a par, in the way of mischief, with the souls of the other world (*Cassumbi*), and with the sorcerers. At times, these spirits enter into the body of some unfortunate, and it is a very expensive matter to turn them out again. On other occasions they play higher pranks, such as swooping down upon a village, and making such a disturbance at night as to allow no one to have any rest, so that the medicine-man has hard work to find a cure and exorcise them.

As there happened to be a diviner in the village, I turned over in my mind whether I might not put him to some account.

I therefore called him apart, made him sundry

presents, showed him very great respect, and pretended to have entire belief in his science.

I then begged him to divine my future fate, a task which he readily accepted, calling together the whole of the inmates of the village, and many of the inhabitants of the Capitão, to be present at the divination.

The ceremony was performed with great circumstance, and he failed not to read in the fragments of the basket, as they were shaken uppermost, the most flattering things concerning me. I was the best of white men past, present and future; my journey was to be crowned with the utmost success, and happiness was to attend all those who went with me.

This prophecy produced the best effect, and no doubt had a great influence over the result of my departure from the Bihé.

I have spoken of the medicine-man and the diviner, and will now say a few words about the sorcerer. The term has a meaning which, though possessing some points of contact with the signification we give to it in Europe, is nevertheless not the same thing.

In South Central Africa any one is, or may become, a sorcerer, and a sorcerer is there understood to be rather a poisoner than a man who has mastery over spirits.

In fact, *sorcery* with these people means poison, and to use sorcery towards any one is *to give poison*, causing sickness, death, or insanity.

This is the rigorous acceptation of the word; but any way the belief in sorcery may occasion an immensity of mischief, and as everything that goes wrong is attributed to sorcery, whether the loss of a skirmish, an epidemic among the cattle, the visitation of storms, &c., it may well be imagined what a wide field is opened for malevolence.

It must not be imagined that there are sorcerers by profession, like medicine-men and diviners. The sorcerer appears as a cause of an effect, and as the cause is at once destroyed, the sorcerer may be likened to a meteor which vanishes almost with its appearance.

Besides these three entities, two of which are definite and the third indefinite, there is yet another pretender, who enjoys a certain importance among these barbarous peoples.

He is the man who calls down and stops the rain. He is one of a class who arrogate to themselves the power of governing the aqueous meteors. Possessing observant minds, these men know, from experience, that with certain winds, in this or that period of the year, it will rain, and that when others prevail it will be dry. And making use of these signs, which are matters of common observance in Europe, and are even recommended to attention by men of science, like Fitzroy and others, they form with tolerable safety their prognostics of the weather, trade on the ignorance of those about them, and claim a power of calling down or staying rain, having previously announced that it will fall or cease.

Poor as the pretensions of these fellows are, they nevertheless impose upon the natives, for, as I have observed, by dint of long experience and careful watching they do not often make mistakes.

These practices that may appear strange in the eyes of most of my readers, were common enough in Europe a couple of centuries ago, and among the poorer classes of our agricultural population it would not be impossible, even at this day, to find them existing.

It is not necessary to go back to the middle ages to meet with royal personages consulting astrologers, and I find that in Portugal a book was printed, *with all the necessary licences*, as recently as 1712, which its author, Gaspar Cardozo de Sequeira, a mathematician of the town of Murça, entitled 'Thesouro de Prudentes' (subsequently added to by Gonçalo Gomes Caldeira, an engineer), which professes to teach the most stupendous and marvellous things to men of culture—be it remembered—inasmuch as the people at large were unable at that period to read. Facts like these should make us charitable in judging the poor blacks of South Africa.

There is a curious law established in the country in respect of women who die in childbirth.

When such an event occurs, it is the duty of the husband to bury his wife by himself alone, he carrying the body on his shoulders to the burial-place and performing all the labour of interment unassisted.

He is then bound to pay the value of her life to her relations, and should he have no means wherewith to do so, he must become their slave. The graves of the mass of the people have nothing to render them distinguishable; the interments are made in any spot in the neighbouring wood that may be deemed suitable.

When I have occasion to speak of the Bihé, I shall dwell more at length upon certain customs prevalent in these countries, and which I had opportunities of studying very minutely, more especially those which refer to the native kings and grandees.

Before leaving the Caquingue I may mention that I found there a custom peculiar to that territory and which is called *supporting women*. When a woman is *enceinte*, a young man will apply to the husband, and ask in marriage the daughter she may bring forth, and if the offer be accepted, the lover will thenceforth be compelled to *support her*, that is to say, supply her with clothing, and satisfy the requirements of her *toilette*.

This custom, as may be believed, is prevalent only among the rich. On the birth of the infant, the future bridegroom redoubles his presents to the mother, and is under obligation to supply the daughter with clothing until the age of puberty, when the marriage comes off. If, however, a son be born, the obligation to clothe mother and son is incurred, and the latter, on arriving at man's estate, becomes the *Quissongo* of his supporter. Later on, I will take occasion to explain the meaning of this term.

Strange as such a custom appears at first sight, it loses much of its extraordinary character upon reflection. In Africa, I met with it in the Caquingue country only; but in Europe I fancy to have observed its prevalence in very many instances, not in form perhaps, but in essence, and bearing in the polished phrase of our drawing-rooms the title of *mariages de convenance*.

Day broke on the 5th of March, 1878, after a most stormy night, in which the rain had come down in torrents. My fever had somewhat abated, but the rheumatic pains were more persistent, and extended from the knees to the ankles. My young negro, Pepeca, was better, so I resolved to start again. Apprehensive, however, of my rheumatism, I hired a hammock and bearers, that were most kindly supplied me by Francisco Gonçalves (Carique). After many cordial adieux, I started northwards at half past ten, and an hour later crossed the little river Cassonge, which runs in a S.E. direction into the Cuchi. It was found to be about 20 feet wide by 6 feet deep. In crossing, my saddle-ox, Bonito, got entangled in some weeds, lost his courage and sank to the bottom. I had great difficulty in saving him, and it was past noon before we could resume our journey.

At 1.15 I crossed another little stream, the Govera,

9 feet wide by 20 inches deep, and at 1.45 camped S.S.W. of the village of Chindua. On my road I had passed near two large villages, Cacurura and Cachota. I had already reached territory that owed obedience to the native king of the Bihé, and found the country all about thickly peopled and well cultivated.

During the night the rain descended in torrents and loud claps of thunder came from the eastward. My fever had completely left me, but the rheumatic pains went on progressing in violence and threatened to extend to my whole body. As soon as it was light, the owner of the bridge over the Cuchi sent to advise me to cross without delay, as a considerable body of natives was advancing on the other side. The bridge was, like most of them, constructed for the passage of men in single file, which, of course, occupies a good deal of time; and it is the law that, when once a party has commenced crossing, any coming in an opposite direction must wait till the bridge is clear. Thanking the messenger for the advice, I broke up my camp immediately, and was able, half an hour later, to get possession of the bridge.

The river Cuchi is at this spot 27 yards wide by 16 feet deep, and runs southward to the Cubango.

One catches a glimpse from the bridge of the magnificent cataract of the Cuchi, rather more than a mile to the north, the roar of which came plainly to the ear.

I stopped for a short time to determine the altitude and then went on towards E.N.E. I passed the little Liapêra rivulet which runs into the Cuchi, and then altering my course to N.N.E., I crossed the Caruci rivulet which flows to the N.E. into the Cuqueima; I there took a noon-day rest in the Charo Woods, situated to the S.W. of the village of Ungundo.

The two rivulets above mentioned, the Liapêra and

the Caruci, mark the separation of the waters which drain into the Cubango and the Cuanza.

The chief of the village of Ungundo, by name Chaquimbaia, paid me a complimentary visit, bringing with him a pig and one or two fowls. I reciprocated the civility and procured from him guides to accompany me on the following morning. During the whole of that day I fell in, on my road, with many bands of armed men, who were on their way to join the forces of the native king of Caquingue, and even after I had camped for the night, a large number of negroes, equipped for war, passed by, bound on the same errand.

Between 7 and 9 P.M. there was a moderate fall of rain, and in the N.E. distant sounds of thunder were audible. The storm came nearer and spread, so that by nine o'clock there were claps of thunder from various points of the horizon, which seemed to be all converging upon my camp, which was situated on a height. At ten, five distinct thunder-claps burst upon us at once, and the most horrible tempest it has ever been my fate to witness broke loose in all its fury. The flashes of lightning succeeded each other with intervals of three to five seconds, and the crash of the thunder was simply incessant.

The air was as yet perfectly calm, and but a few large rain-drops were observable.

The fall in the barometer was scarcely perceptible, and the thermometer maintained a temperature of 16 degrees Centigrade. The magnetic needles lost their polarity, and were in a constant state of oscillation.

One of Duchemin's circular compasses went rapidly round and round.

This state of things lasted until eleven o'clock, when there was another change, even more terrible than before. A wind of excessive violence, in fact a perfect hurricane, came down from the eastward, and in an instant veered from point to point of the compass, until it settled in the S.W. A perfect deluge of rain followed. The wind, in its fury, literally carried our huts into the air from above our heads, and left us thus unsheltered and exposed to the pitiless rain, which fell in torrents until four in the morning, when the tempest began to abate.

No man who has not himself experienced it, can form the slightest conception of what a tempest is at night in the middle of a forest in Central Africa, where, to the reverberations of the thunder, are added the innumerable sounds of the wild denizens of the jungle, answering with discordant cries the voices of the elements at strife.

The rain soon extinguished our fires, the wind carried off the wreckage of the huts, and the lightning in its zig-zag course only served by its momentary brilliancy to show the havoc which the storm had made.

From time to time to the crash of the thunderbolt succeeded another sound, which caused no less alarm. Some giant tree, a very monarch of the woods, which it had taken ages to bring to its state of maturity, was struck to the very heart and went toppling down, destroying others in its fall.

Truly a horrible spectacle, but one fraught with grandeur and sublimity!

Day broke at length, and displayed many a gap in the forest about us, occupied the day before by some magnificent tree, and the earth so soaked with moisture that it yielded water to the tread like a sponge.

The horrors of the night had painfully affected my mind, but they were absorbed, as morning appeared, by my physical suffering. An attack of rheumatism of more than usual intensity affected my every joint, and took from me all power of helping myself. When we started, therefore, at noon I, stretched upon my hammock, had to exercise no common command over myself to stifle in my throat the groans and cries provoked by the intense suffering which the movement of the hammock caused me.

We had not been more than an hour on the road than we found ourselves in an extensive bog where the water came up to the waist-cloths of my bearers.

The earth, soddened by the enormous quantity of rain that had fallen during the night, appeared to be transformed into one vast marsh. We reached some higher ground, after great difficulty, but met with little improvement in our condition as, at 2 P.M., a fresh storm, this time from the eastward, burst upon us. From my hammock, where I was lying a prey to the acutest pain, I encouraged my people to push forward, as I wished to reach the village of Belanga before night.

I have no recollection of anything more till the following day when, awaking as from a trance, I found myself lying in a hut with Verissimo standing by my side. He informed me I was at Belanga, in the village of Vicentes; but I had not the slightest idea either of the road we had come or the night we had passed through, although by my followers' account it must have been a horrible one. I had, in fact, for the time succumbed to fever and delirium.

I found my head somewhat clearer, but my pains acuter, if possible, than before. I could not make the slightest motion, and my very fingers refused to bend. I was fortunate, in this deplorable position, to have such kind hearts about me, for Verissimo and my negroes lavished on me every possible care.

I learned that the river Cuqueima was extraordinarily full, and that to wade across it was simply impossible; but hearing that a small canoe was to be had just below the cataract, I determined to go on and pass the river at that spot.

On reaching the stream, it became necessary to caulk the canoe with moss, for it was a wretched old thing and would barely sustain the weight of a couple of men. The river, swollen with the late rains, was rushing along with great rapidity. After leaping over the rocks which formed the cataract, the waters divided, leaving an islet in the centre, and shortly after they blended again into one channel, some 110 yards wide.

That was the spot selected for crossing. I was laid at the bottom of the canoe with the utmost care, as every involuntary jolt wrung from me a cry of pain.

A skilful boatman handed the paddle, and the canoe left the bank.

The space to be traversed, as I have mentioned above, was scarcely more than 110 yards, but the water was not only made perilous by the rapidity of the current, but by the excessive "choppiness" of the surface caused by the proximity of the falls.

The boatman steered his canoe for the ait, and until he reached the junction of the waters all went right enough; but there the fragile skiff, caught in the furious eddies, could not be persuaded to advance a foot in spite of all the skill and strength of the negro. As I lay, I saw the water leaping in foamy waves about us, becoming larger and more threatening as we got more into the current, and I began to comprehend the extreme peril in which I was placed.

I tried to move one of my arms, but only called forth a groan with the effort. I gave myself up for lost, for if the canoe went to the bottom I was surely incapable of swimming. The canoe, worked upon by the eddies of the seething water, would not go forward, and suddenly the unfortunate skiff began to whirl round itself. My boatman, apprehending we should go to the bottom, determined to jump overboard to lighten the canoe, and warning me of his intention, leaped into the stream.

The canoe, thus lightened, floated certainly higher, but scarcely improved my position, as it was now at the entire mercy of the rushing water.

All of a sudden, a wave leaped over the side and soaked me through. My senses for the time almost forsook me, and I scarcely knew what occurred until I found myself swimming with one arm with all my remaining strength, whilst the other hand was endeavouring to keep from out the water one of the chronometers I happened to have with me.

My sensations returned in the act of swimming, and I remember being conscious of a certain pride in thus buffeting with and overcoming the waves; a task that would have been easy enough to me under ordinary circumstances, as I had been accustomed from childhood to wrestle with the rapids of my native Douro.

The negroes, who are ever ready to admire feats of physical skill, stood upon the bank and animated me with cries of applause.

My pains had ceased, my fever was gone, as if by magic, and I felt, whilst the excitement lasted, as though my strength had returned to me.

When the canoe foundered, out of a hundred men that were present at the spectacle and stood openmouthed and undecided as they looked on, one at least tempted the perils of the waters and leaped in to save me. A less skilful swimmer than myself, he did not reach the bank till after I had done so, nor did he render me any help; but his devotion, at such a time, made a deep impression upon me which will never be effaced. He was one of my own negroes, Garanganja, who, poor fellow, subsequently went out of his mind, unable to bear up against the misery and privations to which we were subjected.

When I got to land I found myself, as I have mentioned, without either pain or fever. I stripped at once; but unfortunately I had no change of clothes, as the whole of the baggage was still on the other side of the stream, so that I was compelled to remain exposed to the burning rays of the sun until they had thoroughly dried my things. The consequence was, the pains and fever came back with redoubled violence, and I remember no more until I found myself next day lying on a bed in the compound of the Annunciada, the late dwelling-place of the trader Guilherme Gonçalves, Verissimo's father.

Racked as I was with pain, and burning with fever, but somewhat better for the long rest, I decided on leaving, so great was my anxiety to meet with my companions.

I started at 11 A.M., and on the road crossed a plain covered with enormous ferns, and observed a number of trees that had been struck by lightning. I saw also a plant that grew abundantly thereabouts and which, if not the gorse that is met with on the lofty mountains to the north of Portugal, is wonderfully like it.

My eyes, that are but little accustomed to that keenness of observation which the study of the vegetable world demands, are not sufficiently tutored to distinguish species, genera and families, unless the difference be strongly marked.

I arrived at Silva Porto's village (Belmonte) at one in the afternoon, and by making a supreme effort reached the house of my late companions.

Confirming verbally what they had told me in writing, they said they had determined to go on alone, and would leave me a third part of the goods and stores, saving such things as were incapable of division, which they would retain themselves. Ivens offered to accompany me back to Benguella, seeing the precarious state of my health, if I made up my mind to return to Europe.

I could but express my gratitude for so generous and disinterested an offer.

CHAPTER VI.

BELMONTE.

In the Bihé—Severe Illness—Improvement—Belmonte—I determine to start for the Upper Zambesi—Letters to the Government—How the Expedition was organised in the Bihé—Difficulties, and how they were overcome—Historical and Social Notes on the Bihé—My Labours—New Difficulties—I leave Belmonte—The Road to the Cuanza—Slavery.

AFTER the twenty days of toil, anxiety, and suffering detailed in the last chapter, I found myself at length in the Bihé—very ill, it is true, but full of faith, and satisfied with what I had done.

Directly after my conversation with my late companions, I left Belmonte and was conveyed in my hammock to the neighbouring village of Magalhães, where, on my arrival, I dropped without strength or motion on to my couch of skins. The first symptoms of inflammation of the brain (meningitis) became perceptible in the pace that the rheumatic pains were increasing in intensity.

On the following day Capello and Ivens came to see me and bring me medicines. I rapidly grew worse till delirium took possession of my senses.

When I recovered consciousness, I thought I was in a dream. I perceived that I was lying on a magnificent bed, divested of my clothes and between fine linen sheets. The bed was upholstered with elegant curtains of pink rep with a snowy white fringe.

I was informed that Capello had come during my delirium and had ordered the bed to be sent me from

Silva Porto's house at Belmonte. I had much ado to believe that an article of such luxury existed at the Bihé.

My attendants had literally covered me with leeches, and the amount of blood they had drawn from me left me in a state of indescribable weakness. The pains had somewhat subsided, but the fever still continued. On the evening of that day the negroes of Novo Redondo waited upon me, and I received them before Magalhães, Verissimo, and Joaquim Guilherme José Gonçalves, eldest brother of Verissimo. The object of their visit was to tell me that they did not wish to go on with my companions, and that they would either follow me or return homewards.

After a great deal of trouble I managed to persuade them to alter their determination and not to leave my friends. I then learned that Capello and Ivens were busily engaged in constructing an encampment, some three miles distance; that they were having all their baggage conveyed thither, and intended shortly quitting Belmonte.

A couple of days later Ivens called on me, and we had a long talk. I gave him all the letters of recommendation with which I had been favoured by Silva Porto in Benguella for the obtaining carriers, and I undertook not to apply to the native king, Quilemo, for any men—thus leaving the field entirely open to himself and Capello. Ivens informed me that they intended moving into their encampment and that they would leave me my share of the baggage in Silva Porto's house. In return I delivered over to him all the loads I had brought with me, together with those under the care of Barros, which had already arrived in safety. Barros himself declared that he had no wish to go any farther, so I dismissed him—as I did also some of the Benguella negroes, who did not care to continue the journey. I wrote a few lines to Pereira de Mello, which the state of my health did not allow me to extend, and then begged to be left alone.

Quite worn-out with such unwonted exertion I was about to turn in to the sheets and seek in sleep a relief from pain and worry, when there rose up before me, like a spectre, a tall, lean man, with cold and impassive look, and strongly-marked features. It was my prisoner, the chief Palanca, the counsellor and friend of the native king of the Dumbo in the Sambo country, whom I had, truth to tell, entirely forgotten.

"Thou hast dealt according to thy will with all thy people," was his greeting. "Some thou hast dismissed and others thou hast retained; what dost thou determine with respect to me; and what is to be my fate?"

"Thou shalt return to thy home," I replied; "thou shalt take back to the Dumbo the gun I promised the king, together with some powder; and thou shalt also take with thee a present for thyself. I owe thee some reparation for the rope put about thy neck at the Cubango, and for the cords with which thy hands and feet were bound."

I then called Verissimo and gave him the necessary instructions for the purpose.

Palanca, as impassible in face of freedom and of guerdons as he had been in that of imprisonment and impending death, retired without a word, and I saw him no more.

The door which let out the grim chief of the Sambo, gave entrance to two other visitors. It was destined that I should have but little rest on the first day of improvement in my health. They were two confidential negroes, Cahinga and Jamba, sent me by Silva Porto

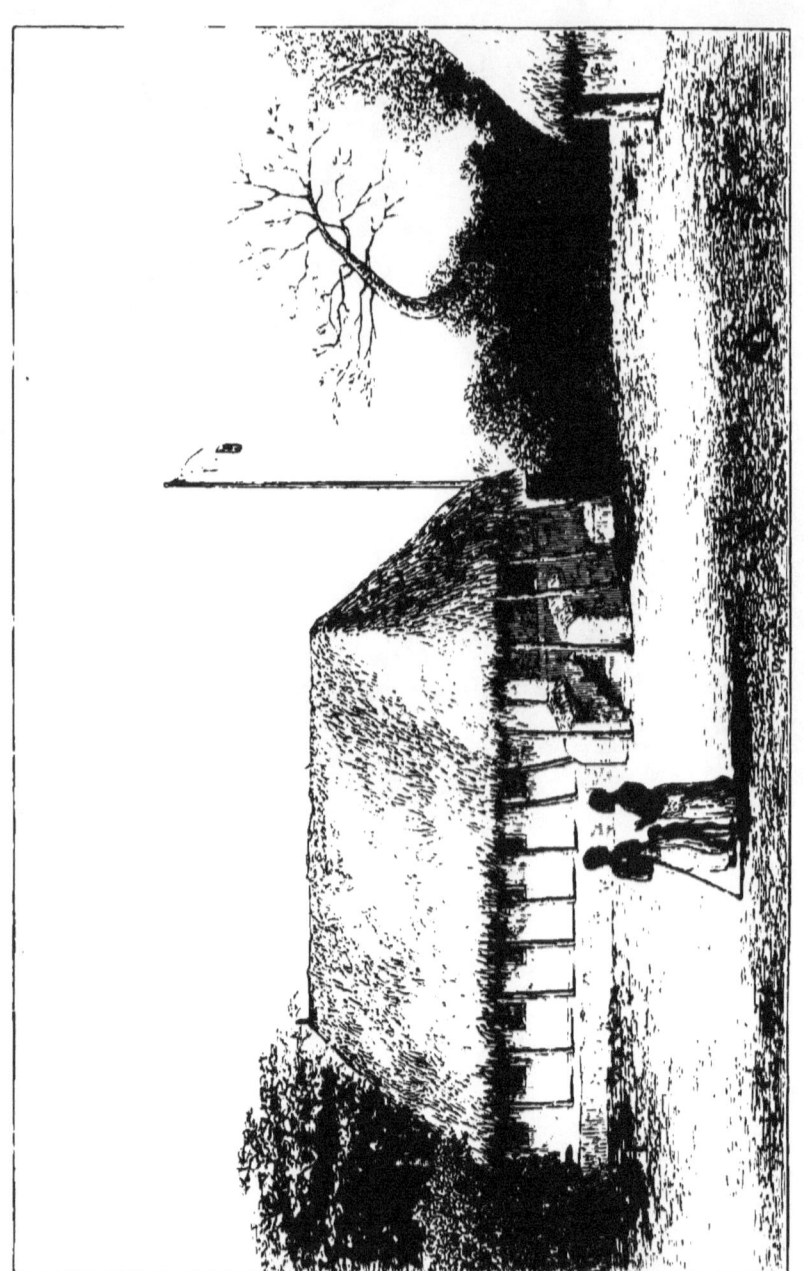

Fig. 16.—BELMONT HOUSE, BIHÉ.

from Benguella. They were profuse in their compliments and offers of service, which, however flattering to my self-love, I could well have dispensed with just then. I got rid of them at last, and with a sentiment of immense relief, found myself between the sheets, and alone!

And yet not quite alone, though the companion left to me would in no way disturb my rest. By my side, the place she best loved to occupy, was the creature that proved my greatest comfort in my journey across Africa. It was Cora, my pet goat, her fore-paws resting on the bed, that, with low bleating, whilst she licked my hands, sought the caresses of which she had been so long deprived.

On the following day Capello and Ivens sent me notice that they were moving out of Silva Porto's house, and in consequence I had myself conveyed thither in my hammock. I found they had left me seven loads of goods, six cases of provisions, a trunk with instruments, and three Snider rifles.

The settlement of Silva Porto, or more correctly speaking the village of Belmonte, is situated upon the highest portion of a rising ground, whose northern declivity slopes gently down to the bed of the river Cuito, which flows eastward into the Cuqueima.

The position of the place is very charming, and from a strategic point of view is strong.

Within its enclosure is an orange orchard, where the trees are ever covered with fruit and blossom, which I found was not the case with any others in the Bihé. This orchard is surrounded by a hedge of rose-bushes, that attain to the height of ten feet and are never without flowers.

Enormous sycamores give shade to the streets and surround the village, which is further defended by a strong wooden stockade.

Under those orange-trees, whose perfumed shade protected me from the burning sun, how many hours, how many days, indeed, did I not spend, pondering over my position, and weaving projects more or less reasonable!

It was there, with my limbs still quivering with pain, and burning with fever, that I conceived and organised in my mind the plan which I was subsequently spared to realise.

If I feel proud of any portion of my journey, the feeling certainly belongs to this particular period.

Fig. 17.—View of the Exterior of the Village of Belmonte in the Bihé.

Later on I frequently jeopardised my life; more than once my boldness took the character of rashness; but it was the thought of my own safety, as I believed, that urged me on. Not so in this place. I was exhausted by disease and had but few resources left me. The road to Benguella and to Europe was open to me, and might be traversed with comparative facility. And as I turned my glances in the opposite direction a thousand difficulties, arising from my separation from my companions, seemed to stand in the way and present a barrier almost impossible to surmount, in regard to any further exploration. And as if this were not enough, the few followers who stood by me appeared to have lost all heart.

And hence arises that feeling of pride and self-

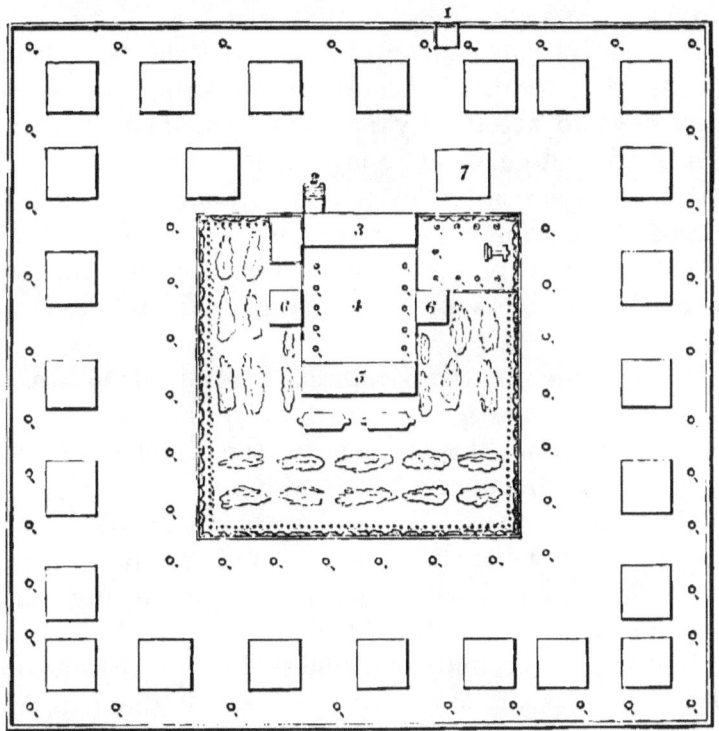

Fig. 18.—Plan of the Village of Belmonte in the Bihé.

o. Sycamores. ═══ Strong wooden stockade. ⌇⌇ Garden Palisade covered with ever-blooming rose-trees. ⁑⁑⁑ Pomegranates. ⌒⌒ Orange-trees. ▭ Gardens. ▯-▯ Cemetery. ▯ Negroes' houses.
1. Entrance of the Village. 2. Entrance into Silva Porto's house. 3. House. 4. Interior *patio* or courtyard. 5. Kitchen and storeroom. 6. Servants' houses. 7. Warehouse.

satisfaction to which I have alluded. For situated as I was, scarcely able to crawl, the determination grew within me *not* to turn my back upon the unknown regions that lay before me, let them be as full of horrors as they might, but to overcome one by one the obstacles as they arose; to reconstruct the edifice I had raised with so much labour and thought, and which had toppled over like a child's house of cards; to create resources where they had no existence; and to organise a fresh and grander expedition out of the ruins of the one which had come to so untimely an end.

My resolve once taken, I lost no time in putting it into execution. I began by engaging Verissimo Gonçalves to accompany me, and managed to make him blindly subservient to my wishes.

After patient study of the direction I proposed to pursue, I determined to make directly for the Upper Zambesi, following the lofty ridge of the country in which the rivers of that part of Africa take their rise.

On arriving at the Zambesi, I resolved to travel eastward and survey the affluents of the left bank of the stream, and descending to the Zumbo, proceed thence to Quillimane by Tete and Senna.

The most experienced traders, who heard of my project, assured me that I should not get half-way to the Zambesi, and I believe they thought me not quite right in my mind to attempt it.

I let them talk, and went quietly on with the organisation of my staff and the preparation of the materials necessary for my plans.

On the 27th day of March, being the first on which I was enabled to use my pen freely, I wrote to the Home Government, and to Pereira de Mello and Silva Porto. I gave them an account of what had occurred up to that time, and begged of them assistance and advice, whilst I submitted my projects to their critical examination. I despatched porters to Benguella with the letters and then went on with my work, feeling daily more confidence in myself.

It will scarcely be credited that a great portion of

the baggage left at Benguella in November, five months previously, had not yet reached my hands!

One morning, shortly after my taking up my temporary abode in the village, there appeared before me the ex-*chefe* of Caconda, and the exiled Domingos, who were on their way back to that little town. They stated that on their arrival at the Bihé they had been engaged by Capello and Ivens in constructing their encampment, and in conveying thither the goods that were stored at Belmonte.

Ensign Castro was considerably depressed in spirit, and out of the cases of provisions left me by Ivens I gave him a supply of sugar, tea, coffee, and other necessaries to help him on his journey.

I fancy that that gentleman, after being the cause of the sufferings I had to undergo, and the fearful risks I had to run, can find no reason to complain of the way I received him at the Bihé, if truth or justice has any place in his heart.

As to Domingos, if I remember rightly, I gave him a letter of recommendation to the governor of Benguella, whom he was desirous of approaching to solicit some favour.

It was in this way I treated the two men who had worked me most evil in Africa, for it was undoubtedly through them I had to face such mortal perils, with inefficient means and ere I had experience to avoid or vanquish them.

At the beginning of April, being then much improved in health, I had sixty carriers in readiness, and only waited for the arrival of the Benguella loads to make up my packages and take a fresh departure.

My life was at that time one of incessant toil; and I was using every leisure moment in compiling a book of notes and data, so as to have at hand the formulæ that were necessary for my calculations. Amongst other things I was making tables of square and cube roots which I calculated for numbers 1 to 1000. I drew up with immense labour sundry trigonometrical forms, for in Europe, in order to render my tables of logarithms more portable, I had had them bound, suppressing the explanatory portions; and through a deplorable oversight, in packing off to Portugal from Loando a quantity of presumably useless baggage, my mathematical books got put up with the rest.

Let not the learned contemptuously smile at my simplicity while recounting the difficulties I had to struggle with in the Bihé, in order to succeed in transcribing upon paper formulæ of so common a kind. A man who is unaccustomed to expound mathematics finds himself not unfrequently at a considerable loss to solve a very simple question, when he cannot lay his hand upon a book which would freshen his dull memory. At the Bihé all my books were wanting, and I therefore set myself to work to supply the deficiency, and whether people may laugh or not, I tell them frankly it was a hard nut to crack. My entire library consisted of three almanacks for 1878, 1879 and 1880, the tables of logarithms I have before referred to, without any explanatory matter whatsoever, the Eurico of Herculano, a volume of poetry of Casimiro d'Abreu, and a little book of Flamarion's, *As Maravilhas Celestes*.

It must be confessed that there was but little to be got out of them to refresh one's memory upon questions of x and y.

But my difficulties did not stop there. I had to do and think about a lot of things at one and the same time,—and things too which were somewhat incompatible with each other. For instance, when I had almost succeeded in reconstructing one of the formulæ of Neper for solving spherical triangles, in would come one of the young niggers to inquire whether the fowl for dinner was to be boiled or roasted. (By-the-bye, during my stay in the Bihé, I consumed one hundred and sixty-nine fowls!) No sooner had I got to work again, after this interruption, than another of the fellows would make his appearance, requesting a bit of soap to wash the linen. He would perhaps be followed by some carriers who wanted specially to speak with me; and I was not unfrequently bothered by envoys from the native chief, whose sole object was to dun me out of some yards of cloth. Truly my patience was often sorely tried.

I had made, and continued making, a great number of meteorological observations.

My chronometers were perfectly regulated and my position determined. Sundry excursions which I made in the country, with my compass in hand, allowed me to draw up a map, a rough one, it is true, but as nearly correct as could be expected or required on a journey of exploration. Notwithstanding all this hard work, or perhaps in consequence of it, my mind was at ease and I gave but scant thought to the tribulations I had to undergo, when I left this quiet shelter of Bihé behind me.

Before resuming the narrative of my adventures, my readers will not deem it amiss if I say a few words about this country, so important and wealthy and yet so little known to us in Portugal, where such knowledge should, nevertheless, be of the highest interest.

The Bihé is bounded on the north by the country of the Andulo; on the N.W. by the Bailundo; on the west by the Moma country; on the S.W. by the Gonzellos of Caquingue; and on the south and east by the free Ganguellas tribes. The river Cuqueima is almost a natural boundary of the Bihé on the west, south and east; but, in point of fact, the authority of the native king of the Bihé extends beyond that river at various points. The country is small in extent, but is thickly peopled for Africa.

I roughly estimated its area at 2500 square miles, and a still rougher calculation made me estimate its population at 95,000 inhabitants, yielding thus barely 38 inhabitants to the square mile; and although this number appears to us very small, as being less than a third of that in our own country, it is considerable for South Central Africa, where the population is, as a rule, very scattered.

Not so very long ago, this territory of the Bihé was covered with dense jungle, abounded in elephants, and boasted but a few sparse hamlets inhabited by the Ganguella race.

The river Cuanza, after its confluence with the Cuqueima, divides the Andulo country from that of the Gamba, which lies to the eastward. The monarch or *Sova* of the Gamba was a certain Bomba, who had a daughter of extreme beauty called Cahanda. This Sova Bomba resided on the left bank of the river Loando, an affluent of the Cuanza.

It happened that the beautiful princess Cahanda requested her father's permission to visit certain relatives, ladies of distinction in the village of Ungundo, the only place of any importance in the Bihé of those days.

King Bomba's daughter having gone on this visit, it also happened that a famous elephant hunter by the name of Bihé, son of the Sova of the Humbe, attended by a numerous suite, passed the Cunene and in the pursuit of his sport reached those remote regions.

One day, this worthy disciple of St. Hubert being hungry, and finding himself near the village of Ungundo, repaired thither to seek materials for a meal. On this occasion he cast eyes upon the beautiful Cahanda, and, as a matter of course, fell deeply in love with her. In questions of love it would not appear that there is much difference between Africa and Europe, and very shortly after the accidental meeting of the young people, Cahanda was wooed and won, and Bihé planted the first stockade of the great village which remains to this day the capital of the country—a country on which he bestowed his own name and whereof he caused himself to be proclaimed the Sova or king. The scattered Ganguella tribes were little by little subjected, and the father of the first Queen of the Bihé, becoming reconciled to his daughter, allowed a considerable emigration of his people to the latter state. The marriage of their sovereign was succeeded by many other unions between the women of the north and the huntsmen who had followed in his train, and thus was the country of the Bihé called into existence.

The Bihenos are therefore Mohumbes,—a name bestowed in the western part of South Africa on the descendants of the Humbe race, who, however, are met with not only in the Bihé, but in various other points, more especially opposite the coast between Mossamedes and Benguella, mixed with the Mundombes, who are the genuine people of that country. At the present date, the true Mohumbe race in the Bihé is represented by what we may style "the nobility" and wealthy inhabitants of the country, descendants of the huntsmen of the first king; but although thus boasting of high lineage, it has greatly degenerated through the admixture of many different races. This is intelligible enough, for as the Bihé, from its very outset, was a great emporium of the slave-trade, and was colonised in great part by slaves of divers races, the lower classes are the issue of an inexplicable mixture, and the nobility itself, by its numerous *amours*, has introduced among its descendants blood of the remotest countries of South Africa.

Of the union of Bihé and the beautiful princess Cahanda was issue an only son, on whom was bestowed the name of Iambi, and who succeeded his father in the government of the country. Iambi had two sons, whereof the elder was called Giraúl and the second Cangombi. Giraúl was proclaimed king on the death of his father, and jealous of his brother's power and influence among the people, caused him to be seized secretly at night, and sold as a slave to a negro who was conveying a gang of such unfortunates to Loando.

By the merest chance Cangombi was purchased at Loando by the Governor-General, whose favourite slave he became. As time rolled on, the tyranny and despotism of Giraúl caused him to be so detested by his people that they conspired against him, and certain of the nobles departed secretly for Loando, laden with ivory, to ransom his brother and set him on the throne after deposing the tyrant.

The then governor of Angola, seeing the profit which might be reaped by the Portuguese crown from this dispute, not only delivered up Cangombi without any ransom, but loaded him with presents, and even lent him assistance in the struggle against his brother. So it came to pass that Cangombi returned to the Bihé with a large following, among whom were many Portuguese. War being declared in due form against Giraúl, he was quickly defeated, being betrayed by the desertion of his men; and Cangombi, more generous than his brother, when assuming the reins of power in his stead, assigned to him a village, with territory attached to it, for his support.

Four years afterwards, Giraúl, untutored by past events, revolted and tried to surprise the capital. Again discomfited and made prisoner, he was delivered by his brother into the hands of the Ganguellas, who dwelt beyond the Cuanza, that *they might eat him*; not that these Ganguellas were positively cannibals, but from time to time they had, it appears, no objection to feast off a fellow-creature.

I did not succeed in learning the name of the governor who lent armed assistance to the younger son of Iambi in order to raise him to power, but I feel convinced that some record of the circumstance must exist among the archives of the Ministry of Marine and Ultramar, as such a step could not fail to be communicated to the authorities of the Home Government.

Cangombi became a great king, and had eight sons, whereof six were reigning Sovas of the Bihé, which is not so surprising when we consider that the nearest in point of kinship to the head of the family assumes the reins of power. Thus so long as there are any sons living of a native king, the grandsons are set aside, and the eldest son of the eldest son only ascends the throne in default of any uncles,—younger brothers of his father.

On account of this law, Cahueue, the eldest son of Cangombi, inherited his dignity, and through successive deaths his brothers Moma, Bandua, Ungulo, Leamula, and Caiangula, did so likewise. The two sons of Cangombi who were not Sovas were Calali and Ochi, they having died early. Ochi came next in order of seniority to Cahueue, and leaving a son, the latter was proclaimed Sova on the death of his uncle Caiangula, as his father's eldest brother left no issue.

This Sova was named Muquinda, and on his death the government passed to his cousin Gubengui, eldest son of the Sova Moma, the nearest of kin to his father. Gubengui was followed in turn by another brother Quilungo, who died, in the act of his proclamation, within his very capital.

Of all the eight sons of Cangombi, only one legitimate descendant remained, son of the Sova Bandua, who then assumed power. This was Quillemo, the reigning potentate of the Bihé.

There nevertheless exists a natural son of Moma, by name Canhamangole, who is pointed out as Quillemo's successor. And as he has many sons, they will in all human probability reign after him.

It will be seen from this brief summary of the history of Bihé that the country is of recent origin, and that almost from its very commencement intimate relations existed between the Portuguese and Bihenos, through the intervention of the Governor-General of Angola—on behalf of the Sova Cangombi, the grandfather of the reigning sovereign Quillemo, and grandson of the founder of the Biheno monarchy.

It happens, therefore, that the Bihé, from the date of its foundation, has been governed by thirteen Sovas in five generations, as represented in the following table :—

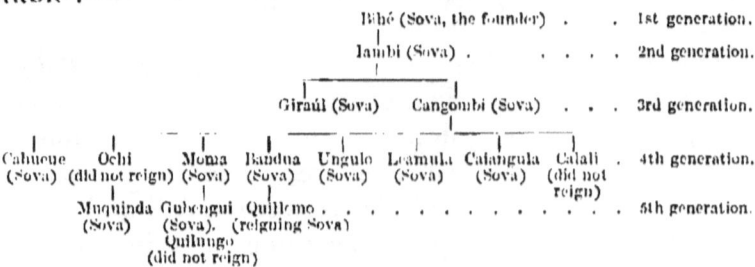

In the map of Angola by Penheiro Furtado the Bihé is marked; but its origin could not long have preceded the production of that map.

The Bihenos are little given to agriculture or to any kind of manual labour. All the work is done by women, who alone cultivate the earth.

The men are fond of travelling, their roaming disposition being probably due to their origin, as their forefathers came from distant parts; and they have no hesitation in penetrating into the most remote regions to carry on their trade in ivory and slaves. Availing themselves of this disposition, certain adventurous spirits, such as Silva Porto, Guilherme, Pernambucano, Ladislao, Magiar and other traders, began to direct and guide the Bihenos in their excursions, and by so doing bestowed a great service upon the world at large, for by opening new markets to trade they opened new fields for civilisation. But it was not their trade alone which little by little increased the commercial activity of Benguella; encouraged by example, and gradually losing their fear of the white men, the natives of remoter

Fig. 19.—WOMAN OF THE BIHÉ, DIGGING.

districts appeared with their wares and did business directly with the commercial houses of Benguella.

The trading excursions into the interior of the country, initiated by the whites, were soon imitated by their black brethren, and at first a few, and afterwards many, obtaining a certain credit in the Benguella markets, proceeded to the Bihé to organise expeditions, which started thence for the interior in search of wax and ivory.

I became acquainted with many negroes who turned over a capital of a thousand to twelve hundred pounds sterling, and some even more; one of them indeed, by name Chaquingunde, originally a slave of Silva Porto, during my sojourn at the Bihé arrived from the interior, where he had traded on his own account to the extent of 14 contos of reis, or about £3500 sterling!

It is not uncommon to fall in with a Portuguese white at the Bihé, who has escaped from the prisons on the coast, acting as secretary to some wealthy negro trader.

Where travelling is concerned as connected with trade, nothing comes amiss to the Bihenos, who seem

Fig. 20.—BIHENO CARRIER ON THE MARCH.

ready for anything. If they only had the power of telling where they had been and describing what they had seen, the geographers of Europe would not have occasion to leave blank great part of the map of South Central Africa.

The Biheno quits his home with the utmost indifference, and bearing a load of sixty-six pounds of goods will start for the interior, where he will remain two, three and four years; and on his return, after that lapse of time, will be received just as though he had been on a journey of as many days.

Silva Porto, whilst engaged in doing business with the Zambesi, was despatching his negroes in other directions, and was trading at the same time in the Mucusso country and in the Lunda and Luapula territories.

The fame of the Bihenos has travelled far and wide, and when Graça attempted his journey to the Matianvo, he first proceeded to the Bihé to procure carriers.

A Biheno rarely deserts his caravan, or makes off with his load—events which are by no means uncommon among the natives of Zanzibar. But the Bihenos have another great advantage over the latter. Although much given to trade in slaves, they do not themselves incite internal wars to procure them; they will purchase them of any who are willing to sell, but they never seek to get them by force. This of course is referable simply to their trade with the interior; for in their wars with neighbouring countries they do pretty much as other negro tribes do, and commit unheard-of cruelties.

Notwithstanding many high qualities, great pluck and readiness to undergo fatigue and danger, the Bihenos have many grave defects; and I do not know in Africa a race more profoundly vicious, more openly depraved, more persistently cruel and more cunningly hypocritical, than they.

These people have a certain emulation among one another as travellers, and I met with many who prided themselves on having gone where no others had ever been, and which they called *discovering new lands*. They are brought up to wandering from their very infancy, and all caravans carry innumerable children, who, with loads proportionate to their strength, accompany their parents or relatives on the longest journeys; hence, it is no uncommon thing to find a young fellow of five-and-twenty who has travelled in the Matianvo, Niangué, Luapula, Zambesi, and Mucusso districts, having commenced his peregrinations at the age of nine years.

A trader who arrives at the Bihé with the intention of pushing into the interior has two means open to him for obtaining carriers. One is to apply to the Sova or the native chiefs for the required number, and make them presents in return; the other to give notice of the journey and wait for the men themselves to apply.

The former is a bad course, for beyond the great expense incurred in the presents that it is absolutely necessary to make to the persons to whom application for the porters is made, the latter are obliged to go, and the party obtaining them becomes responsible for their lives towards their families or lords. And besides, the persons applied to, with the idea of extorting more presents, throw all sorts of petty obstacles in the way so as to retard the departure of the traveller, and one may be sure that their exigencies will increase if the trader be in any way dependent on them.

The second means is far the better, for they who come forward under such circumstances are free blacks; they offer themselves spontaneously, and should any unfortunately die during the trader's service, he becomes, by the law of the country, in no way responsible for the event, inasmuch as the men were under no compulsion in making the offer.

This is a favourable occasion to speak of Quissongos, to whom I alluded in my last chapter, and of Pombeiros. Porters and carriers of whatsoever tribe, Bihenos or not, form themselves into small parties under the command of one among them who becomes their chief. This chief, from the coast as far as

Caquingue, is called Quissongo, and in the Bihé and Bailundo countries, Pombeiro.

It is the Pombeiro who comes forward to negotiate, he having ten, or more, or fewer carriers at his call. The parties or groups are very differently constituted. Some are composed of kinsfolk, who select one of their number to act as Pombeiro, and they are of course all freemen. Others are formed of independent members, freemen also, who combine together under the orders of a Pombeiro in whom they feel confidence; and there are others, consisting of groups of slaves belonging to the very Pombeiros who command them.

The duty of the Pombeiro is to watch over his band, and he is responsible for its members to the head of the caravan. He eats and sleeps with them, and in fact may be looked upon as their captain.

The Pombeiro carries no load, but, in the event of the sickness or death of one of his men, he takes his place as temporary carrier. During the march his place is at the tail of the train, and if a carrier lags behind he is there to look after or assist him.

These men are never paid in advance, and in regular trading journeys their recompense is very small.

For instance, a carrier will receive for the trip from the Bihé to Garanganja (Luapula) twelve pieces of trade cloth to the value of about twelve shillings sterling, and for the return journey a piece of ivory worth say twenty more, making in all thirty-two shillings. This is irrespective of his food, as it is the duty of the chief of the caravan to feed all his people during the journey with the exception of the first three days after leaving the Bihé—the men carrying rations with them for that time.

There is an exception also to this rule. Many traders after leaving the Bihé appoint a certain number of Pombeiros to start for different places, and these fragmentary bands are either detached on their way or at the end of the journey. They entrust to these officials a certain number of loads, for which they are expected to account on their return. These loads are called *banzos*, and the Pombeiro and carriers engaged in such separate ventures board themselves from the very outset of the journey. Saving in this instance, the trader is bound to keep his men and their Pombeiros in food in the manner above described.

The Pombeiros never undertake a venture for any determinate time, and their gains are the same for the shorter as for the longer period. They are employed, in fact, by the job, for it is well known that in Africa the negroes make no account of time.

The customs of the Bihenos are pretty nearly the same as those of the inhabitants of Caquingue, and contact with the whites has produced no change for the better among the natives.

They have no idea of any religious faith, they adore neither sun nor moon, they set up no idols, but live on, quite satisfied with their sorceries and divinations.

Nevertheless, a notion is prevalent among them as to the immortality of the soul, or rather as to its existence in a kind of purgatory until such time as the survivors are enabled to fulfil certain precepts or perform certain acts of vengeance on behalf of the dead.

Their form of government is an absolute monarchy, and has a good deal of feudalism about it.

Every one is, for the most part, a judge in his own cause, and when I speak of the *mucanos* I will describe how justice is done in this part of the world.

The most striking incidents among the Bihenos are those connected with their sovereigns or Sovas, and more especially with regard to the proclamation and death of the latter. Before, however, describing these two great events it is necessary to say a few words about the court.

The Sova is surrounded by a certain number of subjects who are styled *Macotas*, and are assimilated by some to Ministers among ourselves, but this is really not the case. The Macotas form, it is true, a sort of council to which the Sova always submits his resolutions, but of whose opinion he makes but little account. They are *seculos* and favourites of the Sova, but nothing more; and by *seculos* must be understood the nobility, sons of nobles, or personages ennobled by the sovereign.

Many of these *seculos*, who possess *libatas*, or fortified places of residence, assume within their enclosure the airs of native sovereigns, and their people, when addressing them, use the expression *Ná côco*, meaning "Your Majesty."

In addition to the Macotas, there are three negroes who are in attendance on the Sova, and who, when he gives audience, squat upon the ground near him, and carefully gather up the royal spittle, to cast it out of doors. There is another who carries the royal seat or chair, and there is the fool, an indispensable adjunct of the court of every Sova and even of opulent and powerful seculos. To the fool is assigned the duty of cleansing the door of the Sova's house, and the space all round it.

The *libatas* are defended by a strong wooden stockade, almost always covered with enormous sycamores, and a second stockade within the other defends and encloses the residence of the great man. This second enclosure is called the *lombe*.

Having given these brief explanations, I will say a few words as to what occurs on the death or proclamation of the sovereign.

The decease of the Sova is of course known to the

Macotas, who keep the matter a profound secret. They give out to the people that their king is ill, and therefore does not appear. Meanwhile they lay out the corpse on the bed within the hut and cover it with a cloth—at least, this is the custom in Caquingue, but in the Bihé country they hang it up by the neck to the roof of the hut.

The body so remains until putrefaction and insects have left the bones bare; or until, as in the Bihé, the head drops from the body.

It is when this occurs that they announce his death and proceed to the interment of his remains. The bones are placed within an ox-hide and deposited in a hut which exists within the *lombe*, and serves as the mausoleum of all the Sovas. The hut in which the corpse putrefied is demolished and the material of which it is composed is carried out of the enclosure and scattered about the jungle.

From what has been already explained, it is scarcely necessary to say that the death of a Sova is always produced by sorcery or witchcraft, and that some unfortunate has to pay with his life, not for the sorcery, which he never committed, but the private vengeance of one of the Macotas. No sooner is the death of the Sova announced, than the people rush madly about, and for some days not only strip and pilfer all persons who are met with in the neighbourhood of the capital, but make captives of the strangers themselves, and subsequently dispose of them for slaves.

The Macotas then seek out the rightful heir and accompany him to the *Libata grande* or capital; on his arrival, however, he does not at first penetrate the *lombe* or inner enclosure, but takes up his residence among the people, living, for a time, as one of them. No sooner, however, has the heir-apparent entered the Libata, than two bands of huntsmen issue forth, one in search of an antelope (*Catoblepas taurina*), and the other of a human victim.

An antelope being started, a member of the former of the two bands fires at the animal and at once takes to flight, his companions rushing forward to cut off the creature's head, for should this be done by the huntsman who shot it down, he would be at once assassinated and none might say by whose hand.

The other troop, in pursuit of human game, seize the first poor wretch, man or woman, who falls in their way, and hurrying the victim off to the jungle, cut off the head, which they bring back with great care, abandoning the body where it fell. On arriving at the Libata, they wait for the troop on the hunt for the antelope, as it is always much easier to find and kill a man than to find and kill any particular animal.

Having put the two heads into one basket, the medicine-man appears and begins to perform the proper *remedies* to enable the new Sova to assume the reins of government, and his tomfoolery being at an end, he declares that the sovereign may enter the lombe. Attended by the Macotas, the Sova enters accordingly, in the midst of loud acclamations and a great expenditure of gunpowder.

The first step taken by the Sova on attaining to power, is to select from among his women the one he chooses to make his wife, who is styled *Inaculo*; the others still continue to reside in the Lombe but not within the precincts of the royal residence.

Polygamy, however, is an established institution of the Bihé country, as it is of all South Central Africa.

Crimes in the Bihé are always tried in first instance by the parties injured or offended, and it is only if the convicted criminal refuses to submit to the payment of the fine imposed, that the matter, and then only in rare cases, is brought before the Sova. As a rule, sentence is passed and carried out by the injured parties themselves. The word which strikes most terror in the Bihé is *mucano*, a word that does not merely express a crime committed, but an idea that embraces both the crime and the payment of a fine.

All crimes among these people are expiated by money, that is to say, the payment of a fine; and there are no intermediate penalties between a fine and death. When a wealthy person upon whom a *mucano* is pending, refuses to pay, the party injured, if he be powerful, makes a seizure of some of the other's property, for a far higher value than the amount of the fine; and the property so seized remains in deposit, to be subsequently sold, or appropriated by the person effecting the seizure.

Should, however, a seizure be held unjust, the party committing it is compelled by the Sova to make restitution and give a pig, by way of solace, to the party prejudiced.

This system offers a premium to extortion, and not a day passes without the most stupendous *mucanos* being put forward.

One of the most common excuses for its imposition is adultery, wives being urged on by their affectionate husbands to entangle some male friend or acquaintance known to be possessed of means, so that he may be subsequently compelled to pay a *mucano*. The head of a caravan is bound to pay the *mucanos* of his negroes, and he is responsible for their good behaviour.

When a white man, who is liable for the *mucanos* of his negroes, has sufficient force at his command to refuse to make such payment, his accusers will wait—sometimes for years—until they can fall in with another and a weaker white, on whose goods they effect their attachment, letting him know at the same time that they make him the scapegoat of his brother pale-face, out of whom he must get his compensation—if he can.

If a man under the charge of a *mucano* should die, the unfortunate wretch who heedlessly takes up his quarters in the dead man's house, becomes responsible for the former tenant.

The mode in which justice, so called, is administered in the Bihé, is an enormous obstacle to trade, and the source of most serious losses to the Benguella houses.

During my stay in Silva Porto's residence, some negroes came in, bringing with them a hen which they intended using in certain *remedies*, and the gardener, at sight of the fowl, happened to say that it was very like one of his. These unlucky words became the object of a *mucano* and cost the poor gardener some 8 yards of cotton stuff, which he had to pay the owner of the bird.

No sooner does a stranger arrive at the Bihé with goods in his possession, than attempts are made to render him the victim of innumerable *mucanos*, under cover of which great part of his property is filched from him.

The traders on reaching the Bihé are defrauded in this way to such an extent, that in many instances only a third of the goods they have brought with them is left wherewith to do business in the interior. Quilherme the Caudimba, Verissimo's father, on the very last occasion of his going there for trading purposes, was compelled to give up goods to the value of £150 sterling on account of a *mucano* planted on him, through one of his men having purchased a piece of mutton for three cartridges and not paying for it on the same day but offering payment on the day after, when it was refused. During my stay at the Bihé, Silva Porto himself had to pay a *mucano* of £175 on account of even a greater trifle still.

It is this *mucano*, this infamous, because legalised and authorised mode of wholesale robbery, which is the curse of the trade and the main cause of the decline of the Bihé.

It was the *mucano* which drove Silva Porto and all the other honest traders out of the country.

If this were once suppressed, and if the highway to Benguella were rendered safe so that trade caravans might pass to and fro unmolested, we should within an incredibly short space behold the trade of Benguella tripled, and new founts of wealth, now choked and unused through want of security, welling forth and giving life to European industry.

The people of the Bihé are admirably fitted to carry out great undertakings. If we could only eradicate the viper of ignorance which devours their very entrails, raise them from their brute condition to the height of men, and direct them in the right road, we should soon see them take the lead in the march of progress and leave most of the other African peoples far behind them.

The African negroes are not unlike the best breeds of horses, and those among them who at the outset are the most difficult of control end by becoming, with proper training, the most docile and obedient.

The tribes in which indolence and cowardice predominate can with difficulty be civilised: but the laborious and high-spirited would offer a far easier task to their instructors.

The Bihenos, like all the tribes of this part of Africa, are much given to drunkenness. The inevitable aguardente has found its way thither, and where that fails they manufacture *capata*.

Capata, quimbombo or *chimbombo*, for they call the liquor indifferently by the three names, is a species of beer made from Indian corn. In those parts where the hop (*Humulus lupulus*) is cultivated, the people use the conical seeds of that plant wherewith to make their drink.

For this purpose the seeds are reduced to powder, and being mixed with maize flour, the whole is put with a large quantity of water into an enormous pipkin and made to boil for some eight or ten hours. When taken from the fire and allowed to cool, it is *capata*, which is drunk at once.

Acetic fermentation predominates in this preparation, and the alcoholic fermentation is so small that it requires a great quantity to produce intoxication. As the liquor is not filtered, it of course holds a good deal of the flour in suspension, and is therefore rather a fluid mass than a pure liquid. It must have great nourishing power, as there are many of the negroes who will pass a whole day and even more without food, assiduously imbibing *capata*.

In those districts where hops are wanting, their place is supplied by a flour made of maize in a state of germination; the latter produced by burying the corn or steeping it in water for a few days.

In the honey season, considerable alcoholic fermentation is produced by the addition of honey to the *capata*, which becomes, in the course of a few days, transformed into alcohol. The liquor thus prepared is very intoxicating, and it then bears the name of *quiassa*.

There is also another drink which can scarcely be termed refreshing, but is nevertheless both pleasant and very nutritious.

This is made from the root of a herbaceous plant that my imperfect botanical knowledge does not allow me to classify, and which the negroes call *imbundi*. They make a strong decoction of this root which, as containing a great quantity of saccharine matter, ferments readily, and add to it the flour of the Indian corn,—drinking it when cold. This liquor they call *quissangua*.

The food of the Bihé people is almost entirely vegetable, for having little cattle, which they never kill to eat, they go on for months tasting no animal food beyond an occasional treat off the flesh of swine. Pigs abound there in a domesticated state. They were, I believe, introduced by Silva Porto. The country being thickly peopled, game is scarce, and the little there is consists of small antelopes (*Cephalophus mergens*), difficult to bring down on account of their excessive shyness.

It must not be thought, however, that the Bihenos have any objection to flesh; on the contrary, they devour all that falls in their way, and prefer it in a state of putrefaction.

Lions, jackals, hyenas, crocodiles, and all the carnivora are consumed with like gusto, but they have a special liking for dogs, which they fatten up for food. This fondness may perhaps have arisen from the scarcity of animal food existing in the country. They are not positively cannibals, but they do from time to time indulge in a mouthful or two of a roasted neighbour. They prefer, it appears, the old, and a white-haired ancient is a present fit for a Sova or a wealthy native chief who is going to give a banquet.

The sovereigns of the Bihé frequently hold high festival in their *libatas* called the "Feast of the Quissunge," at which are immolated and devoured five persons; viz. one man and four women, who may be thus classified: one woman who makes pipkins; another just delivered of her first child; another who has a goître (a common complaint in the country); and another who makes baskets. The man must be a deer-hunter.

The victims being taken are decapitated and their

heads cast into the jungle. The bodies are brought into the *lombe* or inner enclosure of the royal residence, where they are quartered, and an ox being killed, its flesh is cooked with the human flesh, partly by roasting and partly boiling in *capata;* so that everything which appears at the banquet is mixed with human blood. As soon as this sinister and repugnant meal is ready, the Sova sends out notice that he is about to begin the *Quissunge,* and all the inhabitants of the place hurriedly flock to the entertainment.

The Bihenos, among other strange tastes, are passionately fond of *termites* or white ants, and destroy their habitations to seize and eat them raw.

The people when at home are thorough thieves, and lay their hands upon anything which comes in their way; abroad, however, they not only abstain from pilfering, but, as carriers, are most faithful to their packs.

Should a caravan happen to camp in the Bihé, while passing through the country, notice should at once be given to the chief who owns the land, accompanied by some trifling present; in default of which the inhabitants of the neighbouring village would be authorised to pilfer whatsoever they could lay hands on. The present, however, being made to the land-owner, he becomes at once responsible for anything that is missing.

It is a matter of necessity also to make a present, or rather pay tribute (quibanda), to the Sovereign. It is not advisable to make this offering too costly a one, for his Majesty, as a rule is never satisfied with what is given, but always demands more.

The *libatas* or fortified villages (and they all of them are more or less fortified from the coast to the Bihé), are counterparts of each other, saving such trifling deviations as are due to the configuration of the soil. They are composed of groups of huts constructed of wood and covered with thatch, surrounded by a stockade or palisade, the height of which varies from six to fifteen feet. This palisade is formed of stakes of iron-wood, seven inches in diameter, some of which are merely stuck into the ground, others are secured to cross-pieces by means of withes, whilst others again are strengthened by horizontal pieces fitting into enormous forked uprights.

Another palisade of a similar character surrounds

Fig. 21.

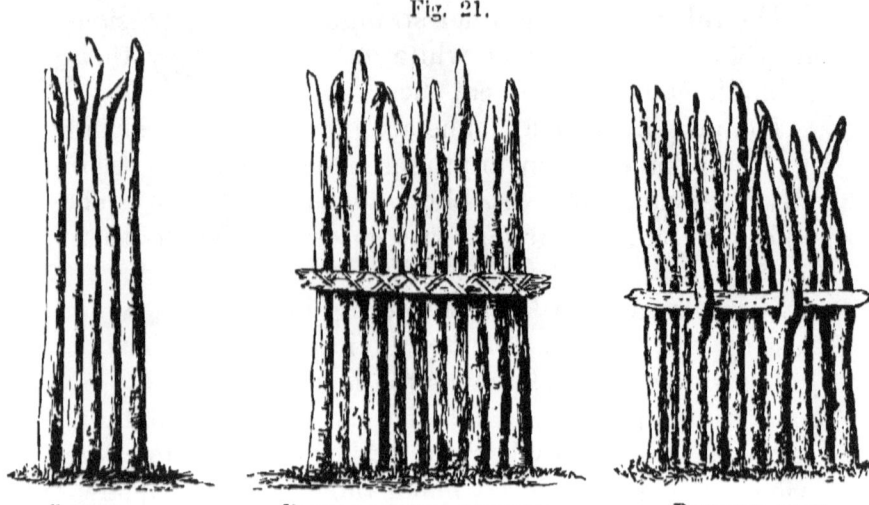

SIMPLE PALISADE. PALISADE BOUND TOGETHER WITH WITHES. PALISADE WITH FORKED UPRIGHTS.

the *lombe* or compound of the chief or sovereign of the place. In many cases I observed groups of houses isolated as it were by means of a palisade.

Most of the *libatas*, and the older ones more especially, are shaded by leafy trees, and are almost invariably on the banks of some river or brook. In many instances they are built over the stream, which thus runs through them.

The majority of them are rectangular in shape, though some are elliptical or circular and others form very irregular polygons. There is not the slightest order observable in the buildings, and the formation of the soil evidently dictates their arrangement.

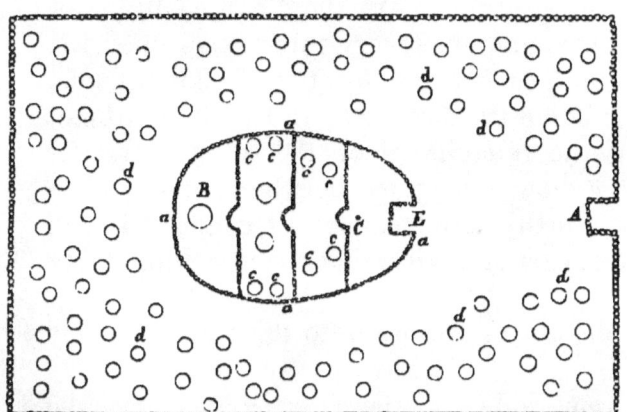

Fig. 22.—Plan of a native Libata or fortified village in the Bihé.

A. Entrance. B. Conical hut where the Sovas are interred. C. Trophy of Horns. *a a a.* Lombe, or residence of the Sova. E. Entrance of the Lombe. o o. Sova's house. *c c c.* Houses of Sova's concubines. *d d d.* Negroe's houses.

Trophy of the chase found in almost all fortified villages.

Fig. 23. — Post erected outside the gate of the vilages.

The villages are fortified to resist the attacks of men, as there are too few wild animals in the district to create any fear of assaults from the latter; indeed, this is so clearly the case that in the interior of the country, where wild beasts abound, the villages are open and unprotected.

Wars among the blacks in this part of the world are, in the majority of instances, utterly causeless, and a reputation for wealth of any particular tribe will be quite sufficient to ensure its being attacked. They are purely freebooting expeditions.

When a sovereign has decided upon a war with another potentate, or tribe, he sends his emissaries round to the native chiefs and *seculos* of the vicinity, to invite them to take part in the campaign; they hasten to the call, and, as was the case in Europe during the feudal times, they come with their warriors to swell the army of their suzerain.

There are some of these people who periodically and systematically make war, and in the Nano country, for instance, they swoop down every three years upon the frontier lands, and carry off the cattle of the Mulonda, Camba and Quillengues districts. Indeed they are apt to boast that the inhabitants of the latter countries breed cattle for them and act as their herdsmen.

It is a noteworthy circumstance as connected with the wars in this part of Africa, that the attacking party is ever the victor.

There are, of course, exceptions to this rule, but they are very rare.

The most remarkable of these exceptions was the attack made by Quillemo, the present Sova of the Bihé, upon the Caquingue country, in which the Bihenos were routed by the Gonzellos, and wherein Quillemo himself became the prisoner of the Sova of Caquingue. He would in all probability have lost his head as well as his freedom, had it not been for Silva Porto and Guilherme José Gonçalves (the Candimba), who paid a heavy ransom for his recovery.

In the wars among the peoples of these countries, perhaps not more than a fifth of the combatants carries fire-arms, the other four-fifths being armed with bows and arrows, hatchets and assegais. A war is looked upon as something great and important, where every man who carries a musket is supplied with thirty rounds of ammunition. The guns in use are those known in the trade as *lazarinas*; they are very long and of small bore. They are manufactured in Belgium, and take their name from a celebrated Portuguese gunmaker who resided in the city of Braga at the beginning of the century, and whose productions acquired very considerable fame both in Portugal and the colonies. His name of Lazaro

—*lazarino*, a native of Braga—is unblushingly engraved on the barrels of the pieces manufactured in Belgium for the blacks—and which are but a clumsy imitation of the perfect weapon turned out by the celebrated Portuguese gunsmith.

The Bihenos do not make use of leaden bullets, which are, they say, too heavy, but manufacture iron ones instead. The cartridges, which they also make, contain fifteen *grammes* of powder and are nine inches in length.

The iron bullets are of much smaller diameter than the ordinary leaden ones, and weigh scarcely six to seven *grammes*. Being of wrought iron, their shape is rather that of an irregular polyhedron than a sphere.

The guns thus loaded, are, as may well be imagined, of but slight precision, and scarcely carry a distance of a hundred yards.

The range of the arrow is from fifty to sixty yards. but in the hands of the blacks it seldom does execution at a greater distance than from twenty-five to thirty yards. The assegais are composed entirely of iron; are short and ornamented with sheep's or goats' hair. They are never thrown—the Biheno in action grasping the weapon tightly in his hand.

I said that the assegai was adorned with *sheep's hair*, and I may mention, while upon the subject, that the sheep in this part of the world have no wool. There are two distinct species existing in the country, which the blacks in Hambundo distinguish by the names of *ongue* and *omême*. The ongue has thick, short hair, and the omême, though furnished with much longer hair, has no pretence to wool.

These animals, of exotic race, degenerate most decidedly from the effects of climate and pasture. The Bihenos have goats of a very inferior race, and their horned cattle are small and of poor and weakly breed.

N 2

Poultry abounds, but, similar to all the domestic animals of the country, the birds are small of body.

Having thus gleaned from my notes what I considered most curious with respect to this interesting country, reserving for a special chapter a fuller account of its climate, capabilities and prospects, I again take up my diary on the 14th of April, 1878.

The rains had been gradually decreasing, falling from six to nine at night only, since the beginning of the month, and yielding scarcely one-eighteenth of an inch of water. The weather was splendid, and even the few flecks of white cloud which after the rains floated for a time at an enormous height in the upper air, at length disappeared to leave the sky perfectly blue and limpid, beautiful by day beneath the rays of a brilliant sun, but infinitely more beautiful at night when sparkling with myriads of stars which shed over this African continent that strangely melancholy light which surely is peculiar to the regions of the tropics.

The weather was admirably fitted for travelling; it was already the 14th of April, and yet I was detained in the Bihé!

The fact was, that I was still waiting for the bulk of the goods and effects left behind in Benguella in the month of November of the previous year, only a portion having reached me at the beginning of March! The delay was becoming a very serious matter. Of the seven bales of goods left me by Capello and Ivens four had already melted away in the maintenance of my Benguella followers and myself.

I had as yet made no present to the reigning chief, who, I feared, would be applying for it, and altogether the prospects of my enterprise looked anything but promising.

I reduced my personal expenses to a minimum, which necessitated my devoting a couple of hours to hunting after game. Of larger game there was none, but on the other hand a good many partridges were to be bagged on the left bank of the river Cuito, on the cultivated grounds belonging to Silva Porto.

I called the spot my "poultry yard," and I went there daily to shoot one or two. I never exceeded that number for fear my supplies should fall short. Somewhat like the gambler who made his livelihood out of the table and retired with just sufficient gains to meet his daily requirements, I had to restrain my sportsman's instincts, and many times tear myself from the field where I might easily have bagged a score of birds. It was not, however, without a struggle that I did so, nor without forcing upon my mind the reflection, that I must not in the mere pursuit of pleasure expend my ammunition, which was getting somewhat low, or destroy the game which represented my future sustenance.

It must not, however, be imagined that Silva Porto's partridges alone furnished my modest table with a dish. Hundreds of African wood-pigeons flitted in and out the shelter of the trees on the banks of the Cuito and in the mornings and evenings came down to wet their beaks in the stream. My young negroes occasionally caught the latter with gins and snares, when they would make a no unwelcome pendant to my *toujours perdrix*, flanked by a dish or two of baked dough, made with maize flour, and which did duty for bread.

In this way I managed to reduce my personal outlay, which represented at least four yards of white calico per day, the cost of a couple of fowls.

The delay, bringing with it, as it did, the rapid decrease of my resources, caused me to modify my plans. The dreaded *mucano* was ever in my thoughts, and I felt that if any one should plant a claim upon me, it would simply render my leaving the Bihé impossible.

Besides this, the want of occupation was beginning to tell prejudicially upon my men, and vices would creep out which amid the fatigue and excitement of travel lay dormant.

Danger therefore, in various shapes, like so many swords of Damocles, hung suspended above my head; so after much cogitation I determined to give myself the advantage of strength, and defend my property at every hazard.

This determination required arms, and not arms alone, but a good store of other munitions of war. I possessed ten Snider rifles, given me by Capello and Ivens; I managed also to obtain eleven of those left behind him by Cameron at the termination of his journey, and to supply these weapons I had four thousand cartridges. Beside these I possessed some twenty flint-lock muskets, some of the last on this system used by the European armies; but I had no ammunition for them. I then made known that I was disposed to purchase all fire-arms considered useless that were brought me. This notice procured me no end of offers which enabled me to pick and choose. I bought those I was able to repair, a matter of no great difficulty to me, as I had learned to become a tolerably good locksmith and gunsmith under the directions of my father, himself a clever mechanician, who still is accustomed to employ his leisure hours in his private work-rooms, which I may truly say are far better fitted up than half those belonging to regular professional artificers.

This explanation reminds me of an amusing anecdote. A gentleman one day, wishing to see my father upon business, came to our villa on the Douro, and hearing a hammering noise in a building not far from the house, directed his steps thither. He found a capacious blacksmith's shop where two men, with arms bare to the elbow, their feet encased in wooden shoes, red nightcaps on their heads, broad leathern aprons hanging from their necks below their waists, their faces and hands black with coal and iron, were hammering at a red-hot bar stretched across an enormous anvil, whilst sparks of fire were flying in every direction from beneath their heavy blows.

The stranger stopped at the door, and inquired: "Is the Doctor within?"

My father, who was one of the smiths, answered him with another query:

"Pray sir, what might you want with him?"

The visitor, a techy person, felt his dignity offended by this seeming familiarity of a mere workman, and rejoined in no very polite terms that he had come to see his Excellency and would not brook, what he considered an insult, from one of his menials.

My father, by his explanation that the blacksmith and the doctor were one and the same person, only made matters worse, for his interlocutor took it as additional insolence, and as both parties were getting very warm, the assistant smith, who was no other than myself, was compelled to come to the rescue, and by explaining matters convince the stranger of our identity.

The circumstance of having been thus accustomed to turn my hand to mechanical work, served me then, as at other times, in good stead, and in fact it might be looked upon as one of the little brooks which helped to swell the river of the happy results of my enterprise.

Another labour was thus added to the many which occupied my days, and I shortly found myself the possessor of twenty-five more guns which the natives had rejected as useless.

Still, ammunition was wanting, and ammunition I must have. I found in Silva Porto's house a complete collection of the *Gazeta de Portugal,* so that there was no want of paper for cartridges. I knew that among the goods I was expecting from Benguella there must be a good deal of powder, and therefore my chief care was to obtain bullets. As to getting any lead, that was impossible, and I consequently soon decided upon making them of wrought iron. Iron was wanting, it was true, but that was not so difficult of attainment.

I again gave out that I was prepared to buy all the old iron offered me, an announcement that was speedily answered by the appearance of a vast quantity of worn-out spades and mattocks, and more especially of hoops of brandy-casks. I had got together some 400 pounds weight before I suspended my purchases.

I then procured four of the country blacksmiths, set up two native forges in the inner court, to the great scandal of the negress Rosa, administratrix of the village of Belmonte, and my own men having produced a lot of charcoal by burning the iron-wood palisade of an abandoned enclosure, we soon commenced operations in right good earnest.

The first labour was to reduce all that mass of iron to cylindrical bars of the proper diameter of the bullets. This the fellows succeeded in doing very dexterously. The hoops were made up into bundles, eight inches long by one and a half inch thick, and being taken from the furnace when red-hot, were plunged into a heap of rubbish and water. On their cooling they were again put in the furnace, and having arrived at the proper temperature they were readily dealt with and reduced into a solid, homogeneous mass. From this point the men's work was easy.

The purchase of the arms and iron had considerably diminished my means, and yet I was far from having everything I wanted. I had no beads—at least, no available ones—for though a bag of them had been sent me by my late companions, they were not current in the districts to which I intended proceeding. I endeavoured to buy some in the Bihé, and with a good deal of trouble managed to procure from the various

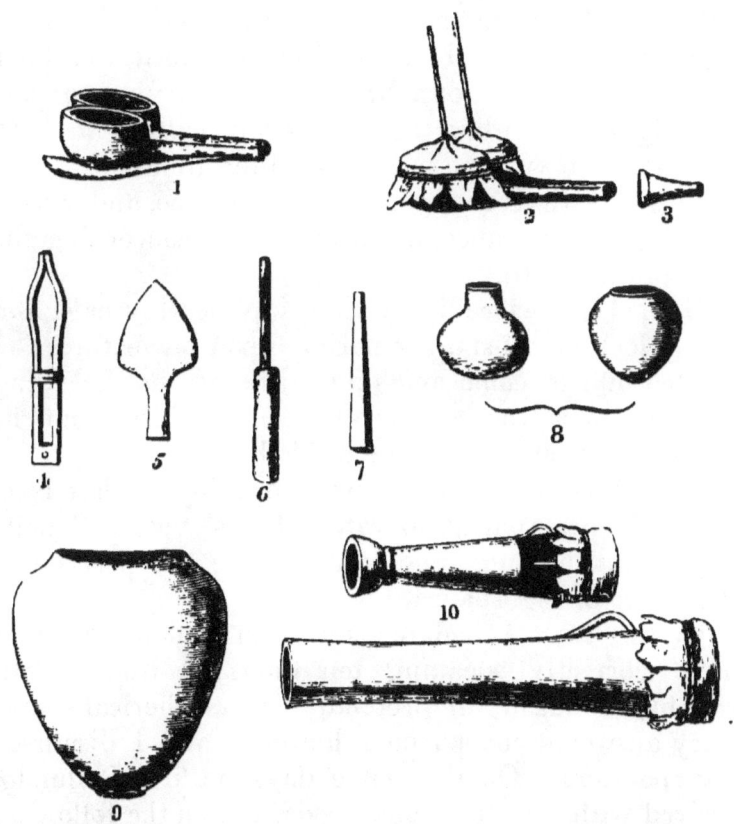

Fig. 24.—Articles manufactured by the Bihenos.

1. Bellows. 2. Bellows ready mounted. 3. Earthenware muzzle. 4. Pincers. 5. Large hammer. 6. A fragment of a musket with a wooden handle used by the smith to remove small pieces from the furnace. 7. Small hammer. 8, Kitchen pots. 9. Large pipkin for capata. 10. Drums.

negroes a small quantity, enough to compose a porter's load. This of course made a fresh hole in my stock of cloth, so that by the 17th of April I had scarcely a pack left.

There was one thing which, since my arrival at the Bihé, I missed exceedingly, and that was an alarum. I had forgotten to bring one with me, and the omission cost me, before my journey was over, very great inconvenience and more than one fever. For instance, whenever I had to make any observations after midnight, I was obliged to keep awake until the hour arrived; and it is not a little dull and trying to pass the night, struggling against sleep, *without a light*, and therefore with no means of killing time.

On the 19th Ivens came to call upon me, and caused me, by his appearance, no little anxiety concerning the state of his health.

He had got exceedingly thin, was deathly pale, and bore a look of constant suffering upon his features. I wanted him to come and dine with me the following day, it being the anniversary of my birth, but he excused himself on the score of his health.

Two days afterwards I went over to my late companions' encampment to return Ivens' visit. Capello was absent, having gone to determine the position of the source of the Cuanza.

By the 25th I had ten thousand bullets ready, or more correctly speaking ten thousand iron pellets, roughly wrought, all pretending to a spherical shape. They answered my purpose, however, and I dismissed my operators. On that same day, the first Bailundos arrived with the Benguella goods, and on the following day more of them appeared. These Bailundos turned out to be insolent fellows, and caused great disorder in Belmonte, indeed the mischief would have assumed larger proportions if I had not myself interfered to check the rioters. I took from out the goods ten packs of cloth, three casks of aguardente, and two bags of cowries.

I still wanted powder and salt, the two things that were yet lagging behind.

I then got ready the present for the Sova, and prepared for my departure, because having the cartridges all in readiness, I knew that I could in two or three days fill them with powder. I sent out messengers to get the carriers together, so as to have everything in a condition to start at a moment's notice.

Fig. 25.

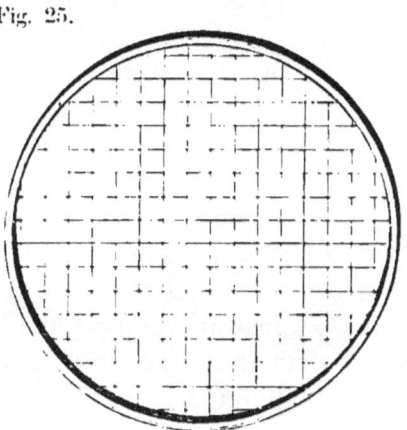

QUINDA, OR STRAW BASKET WHICH WILL HOLD WATER.

LARGE SIEVE FOR DRYING RICE OR MAIZE FLOUR.

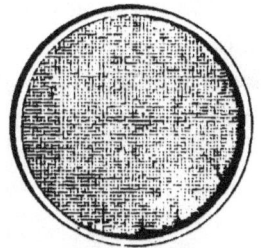

SIFTING SIEVE.

LADLE FOR WATERING THE CAPATA.

On the 29th of April, Silva Porto's blacks robbed me of some trifling article, which made me very angry and threaten to send them back to Benguella. In order to recover my good graces, they came to inform me that they knew where four muskets, which had been stolen from the expedition on the road from Benguella, were now concealed. One of them, it appears, had been appropriated by Mr. Magalhaes, the owner of the premises where I was first quartered in the Bihé. I succeeded in recovering the whole of them.

I was just at this time so busy as to have scarcely a moment to eat my dinner. I had to arrange the loads, and be present at every operation to avoid being robbed, for all the blacks, Silva Porto's and my own into the bargain, were a band of thieves.

There was one exception, however, but one only. This was my negro Augusto, who always displayed the utmost fidelity towards me. When I engaged the porters in Benguella, I hired Augusto among the rest, and at the time attached no sort of importance to him, as there appeared but little to distinguish him from the others, unless it was perhaps his being given somewhat more than they to drunkenness.

In distributing the fire-arms, the men made some difficulty about accepting the Sniders whilst Augusto on the contrary specially asked for one. This first attracted my attention to him. One day, in the Dombe country, I exercised the men in shooting at a mark, and found he was a very tolerable shot. Later on, in Quillengues, I heard that he had asserted among his fellows his determination never to leave me, and as, on account of his herculean strength, and courage, he had secured a great ascendency over his companions, I made him one of my body-guard.

At the time at which my narrative has arrived he had improved his position, and from being a simple carrier was promoted to the rank of chief, a position which he filled most satisfactorily, for those who did not like or respect him, and they were few, were afraid of him.

Augusto was decidedly the best negro I met with in Africa. But no one is perfect in this world, and

Augusto was far from being an exception to the rule. Among his defects I must mention one, which I am nevertheless inclined to treat rather tenderly, for though it is unquestionably a serious failing in an African traveller, it may elsewhere be ranked among the virtues.

To describe it briefly: Augusto was desperately fond of the *fair* sex.

Strong as a buffalo, courageous as a lion, he deemed it, I suppose, his duty to give protection and support to the frail beings he met upon his way.

It would be too long to record his *aventures galantes* from Benguella to the Bihé. Married in Benguella, he took another wife at the Dombe, another at Quillengues, a fresh one at Caconda, wedded anew in the Huambo, and since his arrival at the Bihé had gone through the marriage ceremony three or four times more. He was in fact a true Don Juan, only a black one.

Obedient enough in all things else, he was completely deaf to my admonitions on this subject. But one day, as the complaints of his various wives were loud and troublesome, I summoned him to my presence, reprehended him severely and threatened to turn him adrift if he did not amend. He blubbered a good deal, threw himself on his knees at my feet, made a thousand promises to reform, and said if I would only let him have a piece of cloth to divide among the women and stop their tongues he would have nothing more to say to them, but would remain faithful to his Marcolina, his Benguella partner.

I gave him the cloth and felt delighted at having brought about such sincere repentance.

That very evening, I was disturbed by an unusual noise in a distant part of the village, where songs and other sounds of merriment indicated some festive event.

I had the curiosity to learn the cause and sent some one out to inquire. The reader may conceive my feelings when I was told that it was Augusto celebrating his fresh marriage with a girl from the village of Jamba!

There was no help for it. I saw that this mania of getting married was stronger than his will, and I therefore determined in my own mind to interfere no more with his matrimonial affairs which, after all said and done, compromised no one, as the rascal always kept within the limits of the law.

It was now the second of May, and as yet I had been

Fig. 26.—A Bihé Head-Dress.

unable to get the carriers together, while I was still waiting for the powder and salt that had been despatched from Benguella.

Verissimo was doing his best to collect the men, but hitherto without success.

On the following morning as I was busy about the house, I heard outside, to my astonishment, the sounds of a violin, playing very melodious airs, and totally different to the monotonous music usual among the negroes.

I ordered that the minstrel should be brought in, and there appeared before me a tall, spare, black man, almost naked, with a countenance at once melancholy and expressive.

The instrument he carried was a fiddle manufactured by himself, and out of which he brought sounds as melodious and powerful as could be yielded by a Stradivarius. The body of the instrument, and handle, very similar in shape to those of the European violin, were cut out of a single block of wood, and a thin piece of the same wood formed the top.

It was furnished with three strings of gut, the work also of the musician's hand, and the bow was formed of two similar strings in lieu of the usual horsehair.

It was undoubtedly an imitation of the European fiddle, and not an original instrument.

The wood of which it was composed is called in the country *bóle*, and abounds in the forests of West Central Africa. It might not be amiss to make some experiments with this wood in the manufacture of stringed instruments.

The negro musician sang an air in my praise, *a mezzo petto*, in a most agreeable voice, with an accompaniment of his rude but harmonious violin. He was much applauded by the natives who flocked around him, and I was myself extremely pleased with this unexpected and original music.

Several negroes from the trading station of Andulo arrived at the village, offering for sale some very good tobacco, which is extensively cultivated in that country. It is the Andulo tobacco which the Bihenos purchase and carry to Benguella, where it is sold under the name of Bihé tobacco.

I bought a lot of it, and it cost me, according to my calculations, about a shilling a pound.

By way of curiosity I append the prices of various articles at the Bihé, observing, at the same time, that they are not precisely the same as I was compelled to pay.

A chicken is worth a yard of cloth; six eggs may be had for the same; a two-year-old kid will cost 8 yards; a pig, weighing from 160 to 200 lbs. is valued at one piece of white and one piece of blue cloth, known as *zuarte*; a peck and a half of maize flour may be obtained for 2 yards of cloth, and a like measure of manioc flour for 3 yards. The yards here referred to are the ordinary trade ones, the price of which at the Bihé must never be reckoned higher than 10*d*.

The name given to the trade yard in the Bihé district is a *pano*; two yards are called a *beca*; four a *lençol*; and eight a *quirana*.

The goods proper for the Bihé and commercial marts frequented by the Bihenos are, white cloth; *zuarte*, or blue stuff; printed *zuarte*; handkerchiefs of printed *zuarte*; fine handkerchiefs and checkered ones; striped and other cottons—all of most inferior quality.

The pieces of white cloth contain 28 yards each, and others of better quality 30. The *zuarte* and striped, 18 yards; printed handkerchiefs, 8 yards; checkered, 6; and trade cloth, 12 yards.

Merchandise of good quality is very inconvenient to the traveller in this part of Africa, because whilst it enjoys no greater favour in the eyes of the natives it is considerably heavier.

I had a couple of loads which I had prepared upon the spot, each of which contained 624 yards; and the others, containing scarcely 180 yards of fine white cloth, were much heavier.*

---

* By "a load" I mean as much as a man can conveniently carry, say about seventy-five pounds.—*The Author.*

This will sufficiently prove the inconvenience of the superior material, as in addition to its greater cost, there is increased difficulty and expense in its conveyance: it requires three men to carry it, where one will suffice in the other case.

The argument in favour of the inferior article applies naturally with great force to the explorer, for as he intends to employ his goods in the shape of money, to barter away for the necessaries of life, the same number of yards of common cloth will procure him just as much as a like quantity of fine.

White cotton cloth of inferior quality, and *zuarte*, the blue, are the best money the traveller can carry with him in this part of the world.

The same rule will not hold so good with beads, inasmuch as those which are held in high esteem in some parts will scarcely be looked at elsewhere, in fact, in some cases only a few miles distant; for instance: in the Bailundo country black beads are much sought after, while in the Bihé they are not current at all.

Still, there is one class of beads which is pretty generally received throughout South Central Africa. This is a small red article with a white eye, on which, in Benguella, the trade has bestowed the name of Maria Segunda.

The small cowry is current from beyond the Cuanza to the Zambesi, but the larger kind is of no use at all.

Brass- or copper-wire is esteemed for bracelets, but it should not in these parts ever exceed about an eighth of an inch in thickness.

Scarlet caps, sandalled shoes, soldiers' uniforms, &c., are a delusion and a snare, for though highly appreciated as presents for Sovas and Seculos, they are the very worst of money.

Again, blankets and, above all, those showy rugs used by travellers in Europe, are greatly coveted by the

VOL. I.   O

natives, but must be classified with the uniforms and caps, as forming excellent presents but a poor currency.

This same remark will equally apply to hand-organs, musical-boxes, and other articles of a like nature.

Conjuring tricks and phenomena of physics and chemistry make a certain impression on the natives, but not nearly so deep a one as people in Europe are inclined to think. Not understanding the causes which produce such phenomena, they attribute them without hesitation to sorcery or witchcraft, to which they assign everything they cannot explain.

From my own experience that which produces the greatest impression on the natives, and that which they most admire, is skill in the use of fire arms.

If a man can bring down any prey in presence of an assembly of blacks, if he can put six bullets into a small and distant bull's eye, if he can sever the stalk of a fruit hanging high above his head, or hit a bird upon the wing, he will of a certainty receive great consideration and become for a long time after the subject of conversation.

In proof of this I may mention a little incident which occurred to myself in the village where I was staying. It happened one morning that a Biheno medicine-man made his appearance bringing with him a "remedy" which he asserted was a preservative against bullets.

The belief in such things is general among the Bihenos, and there are many who have been known to expend all they had in the world for the possession of this inestimable medicine, which is supposed to render them more invulnerable than was Achilles of old, inasmuch as there is no possibility of killing *them* even through a single heel.

A civilised Creole, educated in Benguella, whom I fell in with, actually laughed at me when I told him that in spite of any "remedy" to the contrary, I would undertake to put a bullet clean through his body.

But to return to my story. My friend the medicine-man exhibited a pipkin that might have held half-a-pint full of this precious preservative, and asserted that he who took it would become as invulnerable as was the vessel which held the liquid; the best shots in the world, according to his account, having struck it again and again without doing it the slightest injury. In his desire to afford the public an irrefragable proof of his assertion, he had the boldness to defy me to crack the pipkin, taking care, however, to place it at such a distance (eighty paces) as to render it, in his mind, humanly impossible for me to hit so small an object.

I took my rifle and, amid the breathless attention of the assembled blacks, raised it to my shoulder and fired. The pipkin flew into a hundred fragments and the precious liquor spirted far and wide.

Never, surely, was mortal more enthusiastically applauded than I, by the natives there assembled. As for the poor medicine-man, whose anticipated triumph was thus turned into disastrous defeat, he slunk off amid the uproar.

The best of the country marksmen are but mediocre shots, and the arrow and the assegai in the hands of the blacks are much more to be apprehended than firearms.

Verissimo set out to collect the carriers, returning on the 5th May with a few and a promise of others for the following day.

On that same morning I received letters and goods from Benguella, sent me by Pereira de Mello and Silva Porto. The articles, and above all, the kindly words which accompanied them, made a deep impression upon my mind, and I am happy to be able thus publicly to express my sense of their considerate generosity.

Pereira de Mello's parcels contained sixteen muskets, sixty pounds of soap, a watch and a load of salt, all of them articles of the utmost value to me.

But, as I repeat, my gratitude was less awakened towards the worthy Governor of Benguella for this valuable consignment, than for his letter and the expressions of friendship it contained.

Among other things, he said that if I persisted in continuing my journey I might reckon upon the entire support he was able to give me in his official capacity, and that if, perchance, superior orders should restrain him as "the Governor," I might count upon him as "the man" Pereira de Mello.

He further informed me that he had not received from the authorities at home any orders not to furnish me with such means as I might stand in need of; but that should such orders come to hand, he, and the merchants of Benguella, were ready to forward me anything I might require.

Next came Silva Porto's letter, which was no less dear to me.

In it the thoughtful old trader said I must not start without ample resources. That I must apply to Benguella for whatsoever I might judge necessary, and that he would undertake to despatch to the Bihé anything I should ask for.

He concluded his epistle in these terms:—"I am an old man but am still tough and strong: if, my friend, you should find yourself in the interior surrounded by peril, with all but hope gone, try and hold your own, and despatch me a letter through the natives, at any cost. Keep an even mind and wait: for within the shortest possible time I will be with you, and will bring help and means. You know I am not accustomed to make vain promises: if you want me, write, and I will depart forthwith."

No commentary is needed upon words like these, nor, beyond recording them, will I express my feelings of gratitude and appreciation.

These things which I received from Benguella were brought me by a brother of Verissimo's, Joaquim Guilherme, who stated that on the following day the remainder of the loads belonging to the expedition would arrive, and with them the powder I was so anxiously looking for.

As was always the case when a porter came from Benguella, I received a little present from Antonio Ferreira Marques, in the shape of some dainty or other for the table.

On the 6th May, at last, the powder arrived, and I at once set about the great task of filling the cartridges.

During the space of four days I kept between thirty-six and forty men at work at this duty. Every thing was ready by the 10th—and on the 11th May I had collected the whole of my carriers. I distributed their loads, made other preparations, and gave orders for the departure on the following morning.

But when the actual day arrived, and I had every reason to believe I was going to start in good earnest, I discovered there were but thirty men at hand, all the others having taken to flight!

I then learned that on the evening before, a negro, by name Muene-hombo, belonging to Silva Porto, had, with certain other blacks unknown, been among the Bihenos, spreading the report that I intended to lead them to the sea, whence they would never return, as it was my object to sell them for slaves.

Muene-hombo had fled with the Bihenos, and I never set eyes upon him again.

This intelligence caused me infinite depression of spirits.

The carriers whom I had got together at so great a

cost, whom I had hired after the utmost labour and pains, in whose minds I had had to overcome with such care and patience the apprehensions they entertained of my enterprise, had abandoned me after all, under the hasty conviction that I intended to lead them to perdition.

It was a terrible blow.

The news would soon spread throughout the Bihé; the conviction alluded to would shortly take possession of every black in the place: it would override all my arguments to the contrary, and then how would it be possible to get a man to serve me?

I almost myself lost faith in the undertaking, and for the first time after those days in Lisbon when I determined to become an explorer, a feeling of discouragement crept over my mind, for I knew too well how fruitless was the effort to struggle against a conviction of these people.

But who was it that could have induced this fellow, Muene-hombo, to play me so treacherous a trick?

Who were and whence came the negroes who were his companions in the village the day before?

Whose was the hand which pulled the strings in this intrigue?

Again and again did I put these questions to myself without eliciting an answer that was other than a vague suspicion.

All day was I engaged in turning the matter over in my mind. At one time I thought of retracing my steps to Benguella, when the letters of encouragement from Silva Porto and Pereira de Mello suddenly occurred to me.

Why should I not accept the suggestion made me by the former, and beg him to come hither? His presence in the Bihé might procure me followers.

I determined that I would write to him next day,

and the resolution somewhat tranquillised the agitation in my mind.

But with the night came further reflection: before applying to him it was my duty to exhaust every means of obtaining assistance through my own exertions. He ought to be my very last resource.

When day broke on the 13th, I sent Verissimo and certain negroes in the enjoyment of Silva Porto's confidence, to endeavour to contract other men.

They returned, not without hope of success, and the work began anew of organising a fresh band— a labour, as may well be believed, much more difficult than before.

They advised my leaving Belmonte and pitching my camp in the wood at some distance from the village, as they assured me that a caravan upon the march would be more likely to awaken a desire for enlistment.

On the 22nd of May, having succeeded in obtaining a few, a very few carriers, I resolved to make a move with them and my Quimbares on the following day, the 23rd, a determination which I carried into effect by forming an encampment in the Cabir woods.

At dusk of that day, eleven carriers put in an appearance conducted by a negro, Antonio, a man already advanced in years, a native of Pungo Andongo who had been in the service of two renowned traders, Luiz Albino and Guilherme Gonçalves.

The night proved very cold and we were forced to spend greater part of it watching by our fires.

The petty chief of Cabir paid me a visit next day, bringing with him a pig as a present. This civility I returned in kind and we were soon on excellent terms.

He lent me some pestles and mortars and sent some women to make maize-flour.

I walked round his village and passed through the

plantations, where I found women engaged in field-work bent double as they hoed the ground.

On my return to the encampment I was met by a black from Novo Redondo, who had been unable to follow Capello and Ivens on account of the state of his health. He could do little but crawl, and was a prey to a burning fever.

I saw that his condition was hopeless and that his

Fig. 27—Bihé Women pounding Maize.

hold of life was of the meagrest. Still, as he begged me not to abandon him, I had him carried into the camp and placed under the care of Doctor Chacaiombe.

I received a visit from Tiberio José Coimbra, son of Coimbra, Major of the Bihé, who obtained for me a few carriers from among the natives of his village.

In the course of the day some twelve more came in quite unexpectedly, under the leadership of the negro Chaquiçonde, brother of Verissimo's mother.

Hope again began to revive within me, and I set about organising my new caravan.

I determined to make a start on the 27th and to pitch my camp near the dwelling of José Alves, trusting to complete there the number of natives I wanted. I obtained from the petty chief of Cabir a few men to convey thither the loads for which I had no carrier, together with four men and a litter for my Novo Redondo patient.

I was able to leave at the time appointed, stopping, half an hour after we started, in the village of Cuionja, the residence of Tiberio José Coimbra, where an excellent breakfast was awaiting me, with capital tea. There were even table-napkins!

Two hours having been very pleasantly spent I moved onwards, and after four hours' journey reached the village of Caquenha.

I there halted to see old Domingos Chacahanga, the chief man of the place.

Chacahanga, formerly a slave of Silva Porto, was at the head of the celebrated expedition which the latter sent from the Bihé to Mozambique, and which succeeded in reaching Cape Delgado, on the coast of the Indian Ocean; and he was the only survivor of that bold undertaking.

The old man received me very kindly and gave me a young kid.

I had a long talk with him; but all my efforts were vain to elicit from him any reliable data as to his course on that occasion.

That it lay much farther to the north than the indication given on the maps, I had no doubt whatsoever, inasmuch as there were three points which he laid down most clearly.

One was the having, in the Zambesi, left to the southward the country of the Machachas; another, the having crossed the Luapula; and the third, the having skirted the northern part of Lake Nyassa.

Two hours after taking leave of Chacahanga I camped in the Woods of the Commandant, about a mile and a quarter S. E. of José Alves's enclosure.

Night had now fallen and I waited till next day before calling upon this personage, whom Cameron has made so widely known.

It was therefore on the 28th of May that I found myself in presence of this renowned African trader.

José Antonio Alves is a negro *pur sang*, born in Pungo Andongo, who, like many others trading from that place and from Ambaca, knows how to read and write.

In the Bihé they call him a white, because they bestow that name upon every man of colour who wears trousers and sandalled shoes and carries an umbrella.*
In Benguella they condescend to style him a mulatto, of a dark complexion, but the truth is, there is not a drop of European blood in his veins, and he is not only a black in colour and by descent, but has all the instincts of the negro.

He came to the Bihé in 1845, where he was employed by one of the inland traders and subsequently commenced business on his own account, being assisted by Ferramentos of Benguella, now doing a large trade under the firm of J. Ferreira Gonçalves.

José Alves is a man about fifty-eight years of age, somewhat grizzled, thin in body, and suffering from a lung complaint.

He lives like any other black, and has all the customs and beliefs of the untutored natives.

---

* This reminds me of a remark made by Ivens, when speaking of one of these men, in that pleasant way of his which never abandoned him under the most painful circumstances: "I saw," he said, "a jet black negro come into my camp with sandals on, and a parasol in his hand, so I knew he was a white man, and trembled accordingly."

At the time of my arrival at José Alves's house he was engaged in deciding a *mucano*.

In answer to my inquiries I was informed that a mulatto in José Alves's employ had seduced one of the girls belonging to the latter, and as the young fellow had no property of his own, a *mucano* was pronounced upon his mother's family, who *did* possess something and from whom was demanded in payment of the offence an ox or other animal by way of *cleansing his heart*. As he gave me the explanation, the old fellow passed the rugged palm of his huge hand over the part of his trunk which was supposed to contain that organ, and I thus learned that there were ways of dealing with it other than those taught in our European schools.

After the *mucano* had thus been decided, I spoke to him about my journey, which could not, he thought, be carried into effect with the restricted resources at my disposal.

I got him to part with a few beads, but when I broached the subject of carriers, he evaded giving a direct answer by saying that he knew that Capello and Ivens were near the Cuanza struggling against an insufficiency of men; but that if they chose to pay him handsomely there would be no difficulty in arranging matters to their satisfaction. This, of course, was tantamount to saying that if I paid him well, he would let me have them too.

I retired—for the first time pitying Cameron at having been compelled to remain so long in such undesirable company.

I found vegetation in this part of the Bihé very much advanced, and observed in the vicinity of the river Cuito the same termitic disposition of the ground which I described on the banks of the Cutato dos Ganguellas.

As with the carriers who reached me on the 29th,

sent by Verissimo's brother, Joaquim Guilherme, I had sufficient people to proceed upon my journey, I gave orders to start on the following morning.

The powers, however, who preside over mundane affairs had decreed otherwise.

In the afternoon of that day some one or other spread among my men the same reports as were so fatal at Belmonte, and the consequence was that many of them came to me and declared their intention of returning home.

I used all my eloquence to induce them to follow me, but few were inclined to listen.

This was the second time in the Bihé that I was left without people when on the very eve of my departure.

A few Bihenos still remained, and I decided upon getting rid of everything in the way of mere comfort and abandoning all the provisions I had with me, so that with a few more men I might be able to go on.

The difficulty was to get those few more, though I did not despair of the undertaking. A strange adventure which occurred on the 30th helped to crown my hopes with success.

A lot of loose characters and deserters, who had escaped from the military stations on the coast, suddenly appeared in the Bihé.

One of these worthy, or unworthy citizens called upon me and pronounced a set speech, which, on account of the profuse employment of the first consonant in lieu of the seventeenth, and repeated use of terms only used in my own province, betrayed him as a fellow-countryman.

Even if the style of the discourse had not been that of a consummate rogue, its essence would have sufficiently stamped the orator as a villain, with a soul no better than a sink of rottenness which exhaled, with every phrase, more deadly poison than any fetid marsh of that tropical clime.

After counselling me to use the arms and ammunition at my disposal in a most villainous undertaking, to which he did me the honour of offering himself as an associate, he terminated by saying that if I refused his terms, he would at any cost employ the influence which he possessed over the natives to compel them to abandon me and thus render it impossible for me to take another step in advance.

At the close of this peroration, which my man considered would be a triumphant argument to secure my decision, he demanded an immediate reply.

I did not keep him long waiting. Calling my Quimbares I ordered them to seize and tie him up to the first tree, and then caused to be administered to the rascal some fifty lashes, that we might become better acquainted with each other, for though I knew him thoroughly before he had spoken a dozen words, he had not had the same opportunity till then of knowing *me*.

After his flagellation, I made him a little speech in return for his harangue, wherein I told him he must consider himself my prisoner during the time we stayed in the Bihé, and that his daily ration of food should be accompanied by an equal dose of the lash if he attempted to escape.

I then called all my people about me, and pointed out to them that the heart of that white man was blacker than the skin of any bystander.

The news of this act of justice spread like wildfire in the villages all about, and raised me immensely in the estimation of the negroes, on whose fears the fellow had already begun to trade.

On the following morning, Sunday, Pombeiros of the vicinity came to offer me carriers, whom they promised to produce within three days.

These promises were again and again made, but no carriers were forthcoming, so that by the 5th of June, being reduced almost to despair, I determined upon abandoning a lot of baggage and going on with the remainder.

With this view I called my Pombeiros together and communicated to them my decision.

We held a long council in which I maintained my determination, and gave orders for the carriers to accompany me to the river Cuito with the baggage I had decided to part with, in order to cast it into the stream.

This resolution was in fact about to be carried into execution when Dr. Chacaiombe put in his word and begged me to defer the fulfilment of my project for a few days; he further advised me to hire a certain number of men in the neighbouring hamlets to transport the whole of the goods to the Cuanza, and that meanwhile he would make an effort to get what was needed through a Sova, a friend of his, and would meet me on the banks of that river.

This advice having been duly discussed and adopted, I decided upon starting on the 6th and remaining till the 14th by the Cuanza; this would allow Chacaiombe eight clear days, beyond which, as I assured him, I could not possibly wait.

My Pombeiros displayed the utmost devotion, and upon a proposal of Miguel's (the elephant-hunter) they all resolved to shoulder loads themselves, although this was not only contrary to usage, but inconvenient upon the march, where they have their own special duties to attend to.

Having obtained a number of men on hire, I made all preparations for immediate departure. At the close of that day my poor patient, the Novo Redondo man whom I had succoured in Cabir, sunk under his disease.

At nine o'clock in the morning of the 6th of June I broke up my camp, having a lot of natives for temporary carriers, hired at the rate of a yard of cloth per day.

I travelled eastward, and two hours later camped near the village of Cassamba.

This place is nestled in the midst of an extensive and dense forest which seemed a likely place for game, but where I only succeeded in bringing down a few guinea-fowl.

On starting the next morning, the 7th, I was met by the chief of the Cassamba, who came to pay me his compliments and offer me an ox as a present.

I excused myself for not making a suitable acknowledgment of his civility then and there, on the ground of my carriers being on the march, but begged him to send some of his followers to my new encampment to receive it at my hands.

After three hours' tramp, having during the two last traversed extensive marshy plains, I arrived at the left bank of the river Cuqueima which there ran northwards, being eighty-seven yards wide and ten feet deep, with a current running at the rate of thirteen yards a minute. I fitted up my mackintosh boat and succeeded, though with vast trouble and delay, in effecting a safe passage to the other side with all the goods and men; it was a great achievement of its kind, for the little skiff would only carry five persons, although the floating power of its air-chest was considerably superior.

The passage being effected and finding myself on the right bank on marshy soil, I sent to beg the Sova of the Gando to allow me the use of some huts for the shelter of my people during the night.

He came out himself to see me and to place at my disposal the *lombe* of his village, which I accepted and where I at once took up my lodging.

Shortly after, several negroes made their appearance

sent by the chief of Cassamba to receive the present I had promised him, bringing with them as a token the assegai of their ruler, which I had seen in his hand that morning.

It is a custom among these people where written language is unknown to send some known object by the bearer of a message, to prevent the possibility of doubt in respect of the sender.

I of course returned by them the promised gift.

I had a long talk with the Sova Iumbi of the Gando, who was lost in wonderment at everything I had about me. He gave me a splendid ox and was made happy in

Fig. 28.—GANGUELLA, LUIMBA AND LOENA WOMEN—METHOD OF SHAPING THE INCISORS.

return with a piece of striped cloth and a few charges of gunpowder.

Early the following morning we were again afoot, and two hours afterwards camped about a mile to the west of the village of Muzinda.

Before leaving I ordered my white prisoner to be unbound and landed on the opposite bank of the river, for having crossed the Cuqueima and consequently being out of the Bihé territory, it was impossible that he could do me any harm.

Several women from the village of Muzinda came to

my encampment; some among them had their faces painted green, there being two transverse stripes across the head from ear to ear and two others descending from them, crossing each other between the eyes, passing along each side the nose and being connected by another, traced above the upper lip.

The head-dresses of these Ganguella women are wonderful to behold, many at a certain distance looking like the bonnet of a European lady.

The whole of the men I saw had the two front incisors of the upper jaw cut into a triangular shape, thus forming a triangular aperture with the vertex turned towards the gum. This operation is performed with a knife, which is struck by repeated, slight blows.

One of the natives gave me a sugar-cane six and a half feet long by half an inch in diameter, and assured me that the plant was abundant in the neighbourhood.

During our stay a small caravan left Muzinda bound for the regions beyond the Cuanza to procure wax in exchange for dried fish of the Cuqueima.

The natives composing the party were almost naked, their only covering being two scanty skins hanging from a narrow leathern belt.

The women were even more scantily supplied with clothing!

I had a visit from the petty chief of Muzinda, who brought me an ox as a present, which I returned in the same way as I had done with the Sova Iumbi of the Gando.

On the 9th of June, I camped on the left bank of the river Cuanza, E. N. E. of the village of Liuica. At that point the Cuanza is a less considerable stream than the Cuqueima, as its width is 55 yards by 6 feet deep, with a current of 16 yards per minute.

Its bed is composed of fine white sand, and the transparency of its waters is noteworthy.

The river winds its serpentine course through a vast plain from a mile and a half to two miles broad, enclosed on either side by gentle green slopes clothed with trees.

The plain itself appeared covered with excessively tall grass and reeds, so thick and stiff that it was difficult to make a passage through. The soil was more or less marshy.

As I had to wait there some five days, as agreed with Dr. Chacaiombe, I at once, on my arrival, ordered a far larger encampment to be constructed than I usually built for merely one night's shelter.

The Sova of Quipembe was the first to pay me a visit. All the petty chiefs between the Cuqueima and Cuanza are subject to him, and he is himself a tributary to the Sova of the Bihé. His subjection is, however, merely nominal, as he entertains no fear of being attacked on account of the facility of defending the line of the Cuqueima, and because the greater portion if not all the boats upon the stream are owned by the Ganguellas.

He brought me a sheep as a gift, excusing himself for not presenting me with an ox, on the ground of his village being at so great a distance.

I had also a visit from the petty chief of Liuica, who offered me an ox.

This chief, a man of comely face and figure, became quite an *habitué* of the camp during my stay in the neighbourhood.

One day, when he had been watching me fire at a mark and admiring the precision of the aim, his great herd of oxen happened to pass that way.

I proposed to him, laughingly, that he should give me an ox if my young black attendant, Pepeca, could kill it with a bullet.

He looked at the lad and gave his consent.

Pepeca, who was a very tolerable shot, having been

taught by myself, took his rifle and aiming at a fine beast that was somewhat separated from the herd, brought it down with the blow. The Ganguellas were perfectly thunderstruck; the chief however was as good as his word, though he had evidently expected a different termination to the affair. He merely requested me to let him have the skin and a mouthful of the meat, and gave me up the animal.

The Ganguellas between the Cuqueima and Cuanza are of a different race to the other tribes bearing the same name; near the Cuqueima they are called Luimbas, and near the Cuanza, Loenas.

On the 12th of this month there occurred an extraordinary adventure, which I cannot refrain from recording here.

I was leaving the camp for a stroll when some of my negroes came up to me, accompanied by a mulatto, who was a perfect stranger; they introduced him as the chief of a caravan, who begged my permission to accompany me some distance on the road I was travelling, and allow him meanwhile to take up his quarters in my encampment, to secure his safety.

I consented to his request, although I own it was rather against the grain.

That same night I remained up later than usual talking with my Pombeiros, and seated at the door of my hut we discussed the probabilities of Dr. Chacaiombe's success in his undertaking, when I heard a singular noise in one corner of the camp.

It was as like as possible to the sound of a hammer on an anvil; and my curiosity being awakened I despatched my henchman Augusto to discover the cause.

He returned after a few minutes with news that in the part of the encampment occupied by the Biheno mulatto who had asked me for shelter, there was a gang of slaves, arrived that very evening from the Bihé.

All my people were then asleep in their huts with the exception of the three or four Pombeiros who were keeping me company.

I restrained my anger, which for a moment had almost got the better of me, and summoned my uninvited guest to my presence.

He appeared at once and seated himself near the fireplace in front of me.

I asked him what was the meaning of that clanking sound of iron, to which he replied with the utmost effrontery that they were chaining up some *kids* which he was conveying into the interior for sale.

And so, in my own encampment, upon which floated the Portuguese flag, there was actually a gang of slaves!

Keeping myself as cool as my nature would permit, I let the fellow know that he was telling me a lie, and bade him forthwith to knock off the chains of the unfortunates he had with him, and deliver them over to me, free.

This he not only refused to do, but received my command with a grin of contempt.

I then lost all patience, and my rage, which had been kept down with immense difficulty from the moment I learned the character of my guest, now boiled over.

I made a dash at the fellow, seized him by the throat and drew my knife with the intention of plunging it into his body, when I became sensible that the muzzles of two or three guns were within a foot of his head, and were on the point of being fired by my attendants. This brought me to my senses, and whereas the moment before I would have killed the wretch without hesitation, I now used my efforts to save his life.

The hubbub occasioned by this affair woke up all my men, who came rushing to the spot, and a cry arose to exterminate the whole Biheno caravan.

Knowing the ferocity of the negroes when they feel they have strength on their side, I began to be alarmed for the lives of the innocent who might be sacrificed with the guilty.

Of course, with the exception of the Pombeiros who had been with me since the commencement of the scene, all were ignorant of the cause of the uproar. They saw that I was in a tremendous rage, and they became clamorous for a victim.

I succeeded at last in quelling the tumult and in obtaining a hearing.

I then ordered Augusto to set the slaves free and bring them before me, together with all the cords and shackles they could discover in the huts where the poor creatures were confined.

The shackles were all cast into the Cuanza with the exception of those which I reserved to bind the blacks who had acted as guards over the poor slaves.

As to the slaves themselves, I told them they might go wheresoever they pleased, and that I would take care that the guards should remain bound long enough to prevent the possibility of their overtaking their late prisoners. They disappeared in a twinkling with the exception of one young girl who begged to be allowed to remain with me, as she did not know where to go. I may mention that it was not till I broke up my encampment that I set at liberty the leaders and guards of that gang of slaves.

The 13th of June came and passed without any news of my Doctor, and in the evening of that day I distributed such loads as I was able to do, about eighty-seven, which I afterwards, with infinite reluctance, reduced by twelve, and made one heap of those which were irremediably condemned.

It may be credited that the choice was one of no little difficulty, and I fancy that one of the hardest nuts that an explorer has to crack is to choose between such goods as are positively indispensable and those he can part with. If not a more difficult, it is at least as stiff a problem as to discover the mode of determining a good longitude.

I made up my mind to abandon whatsoever had only convenience or comfort to recommend it; to reject everything in the shape of comestibles for my personal use, together with part of those I carried for my people, and several loads of beads given up to me by my late companions, and which, being purchased in Loanda, were of problematical value in the interior of the country where I proposed to penetrate.

If on the morning of the 14th there was no news of Chacaiombe, the condemned loads were to be destroyed. some by fire and others by sinking in the Cuanza.

"And why," my readers may perhaps inquire, "should they be so destroyed?"

Because the chief of a caravan on his march through the interior of Africa, where he has to employ carriers, is bound to destroy and render useless all articles he may be forced to abandon, and this for two reasons—one out of respect for his own people; and the other, for the natives of the districts he is passing through.

If he once consented that his own carriers should appropriate as their property any portion of abandoned goods, there would be a daily falling out of the ranks of porters, on the plea of illness, as an excuse for making off with what they had thus acquired, and a perfect system of robbery and faithlessness would be inaugurated.

On the other hand, if the natives of the country came to learn that goods were left behind for want of men to carry them, they would not fail to ply the porters of any future caravans with unlimited *capata*, or other intoxicating drink, so as to incapacitate the men and compel the chief of the troop to abandon his property, which they would not dream of doing if they reaped no advantage from it, through the system of destroying all goods which cannot be carried on.

This was a lesson taught me by Silva Porto, and one that I constantly put in practice.

It fell out therefore that when the 14th day arrived, and no intelligence came to hand of Dr. Chacaiombe, I destroyed sixty-one of my loads!

RAPID RETROSPECTIVE GLANCE.

The annexed Map shows my track from Benguella to the Bihé.

I have endeavoured to furnish it with all such details as it was possible, in a journey of exploration, to collect in the shape of geographical and topographical data.

Many of the places marked were determined astronomically, and the intermediate ones were found roughly by the points of the compass and projection of the distances gone over, distances that were obtained by the pedometers and the time employed in traversing them.

The positions of Benguella, Dombe, Quillengues, Ngola and Caconda, which I have given on the map were fixed by Capello and Ivens, and as I barely obtained the results of the calculations, I adopted the data as furnished me by Ivens without the initial observations. From Caconda to the river Cuanza, the positions astronomically determined by myself appear preceded by such initial observations.

Result of the Observations of Capello and Ivens from the Coast to Caconda.

Names of Places.	Longitude E. of Greenwich.	Latitude S.	Deflection of the Needle.	Inclination of the Needle.	Altitude in feet.
	° ′ ″	° ′ ″	° ′	° ′	
Benguella . .	13 25 20	12 34 17	23 30 W.	39 37	23
Dombe Grande .	13 7 45	12 55 12	23 26	39 44	321
Quillengues . .	14 5 3	14 3 10	23 3	40 40	2,788
Ngola . . .	14 39 1	14 16 46	. .	. .	4,506
Caconda . . .	15 1 51	13 44 0	22 30	. .	5,507

Having separated from my companions in Caconda I went on with the work we had commenced together, but was unable to make any observations of inclinometer and magnetic force, because the only instruments we brought out with us for this purpose remained in possession of Capello.

I shall begin the recital of my labours by the determination of the geographical co-ordinates from Caconda to the left bank of the Cuanza, to which point the narrative, closed by the foregoing Chapter, extends.

In the following Table I have endeavoured to condense the necessary data with a view to verifying the results I set down.

The whole of these observations, calculated in Africa, were recalculated in London by Mr. Selwyn Sugden, 1st Lieutenant of the Royal Navy of Great Britain.

TABLE OF THE ASTRONOMICAL OBSERVATIONS MADE BY MAJOR SERPA PINTO BETWEEN CACONDA AND THE RIVER CUANZA.

Year 1878.	Where Observations were made.	Time by Chronometers.	Greenwich Time.	Nature of Observation.	App. height of Star.	Latitude South.	Longitude in Time.	Error of Instrument.	No. of Obs.	Results.
		h. m. s.	h. m. s.		° ′ ″	° ′	h. m.	″		° ′ ″
January 14	Vicete (near the Cunene)	8 10 24	+1 0 15	Alt. Mer. ♄	101 3 0	14 2	...	−3 30	1	Latitude ... 14 2 S.
,, 16	Feude (Cunene)	10 27 44	+3 23 2	Chron. ☉	101 2 0	,,	...	,,	1	Longitude ... 15 14 E.
February 12	Libata of Palanca	5 10 2	+3 23 16	,,	104 31 0	,,	...	−0 50	1	Lat. ... 15 25 E.
,, ,,	,,	7 55 0	−1 0 0	Alt. Mer. ♄	97 5 10	13 20	...	,,	1	Long. ... 13 20 S.
,, 15	Libata of Capoco	10 30 56	+3 27 18	Chron. ☉	99 6 30	,,	...	,,	1	Lat. ... 15 27 E.
,, ,,	,,	9 3 0	−1 0 0	Alt. Mer. ♄	98 30 30	13 9	...	,,	1	Long. ... 13 9 S.
,, 18	,,	9 57 15	+3 27 27	Chron. ☉	115 5 30	,,	...	,,	1	Long. ... 15 30 E.
March 16	Belmonte (Bihé)	10 18 14	+3 28 8	Alt. Mer. ♄	104 15 30	,,	...	,,	1	Lat. ... 15 28 E.
,, 18	,,	10 25 0	−1 4 0	Chron. ☉	131 38 30	12 22	...	,,	1	Lat. ... 12 22 S.
,, ,,	,,	5 6 10	+3 31 43	,,	104 58 40	,,	...	,,	1	Long. ... 16 51 E.
,, 22	,,	{5 3 1 / 9 31 41}	...	Equal alt.	103 21 10	,,	...	,,	2	Time ... 3h 31m 54s
April 2	,,	...	...	Alt. Mer. ☉	144 49 0	,,	1 8	−3 30	1	Lat. ... 12 23 S.
,, 3	,,	...	...	,,	144 4 0	,,	,,	,,	1	,, ... 12 23 S.
,, 4	,,	...	...	,,	143 20 0	,,	,,	,,	1	,, ... 12 23 S.
,, 5	,,	...	...	,,	142 32 0	,,	,,	,,	1	,, ... 12 23 S.
,, 6	,,	4 53 40	+3 34 29	Azimuth 266-30 ☉	93 34 30	,,	,,	−1 0	1	Variation ... 21 11 W.
,, 7	,,	...	...	Alt. Mer. ☉	141 47 0	12 22	1 8	−3 30	1	Lat. ... 12 22 S.
,, ,,	,,	...	...	,,	141 3 0	,,	,,	,,	1	,, ... 12 22 S.
,, ,,	,,	0 8 32	−0 55 43	Alt. prox. of Mer. ☉	140 14 0	,,	,,	,,	1	,, ... 12 22 S.
,, ,,	,,	10 50 54	...	Eclipse of 1st satellite of Jup.	...	...	...	...	...	Long. ... 16 46 E.
,, ,,	,,	10 55 6	+3 34 54	Chron. ☉ ♄	65 48 0	,,	,,	−1 0	1	Diff. for place 4h 42m 23s
,, 23	,,	9 4 23	...	Eclipse of 1st sat. of Jup.	...	...	...	...	...	Long. ... 16 49 E.
May 24	Woods of Cabir (Bihé)	9 38 16	+3 37 26	Chron. ♄	71 30 40	,,	1 7	−0 30	1	Retard. 4h 44m 54s
,, 31	Woods of Commandant	9 38 55	+3 42 47	Alt. Mer. ☉	113 10 40	12 22	...	−1 25	1	Lat. ... 12 22 S.
,, ,,	,,	9 12 5	+3 43 56	Chron. ☉	79 22 50	12 28	...	,,	3	Long. ... 16 53 E.
June 1	Liuica (banks of Cuanza)	...	...	Alt. Mer. ☉	86 38 10	,,	...	,,	1	Lat. ... 17 9 E.
,, 9	,,	6 22 33	+3 45 52	Chron. 2	110 26 40	,,	...	,,	1	,, ... 12 28 S.
,, 10	,,	6 4 53	...	Eclipse of 2nd sat. of Jup.	63 59 30	12 35	1 9	−0 35	1	Diff. for place 4h 54m 34s
,, ,,	,,	...	...	Alt. Mer. ☉	108 15 20	,,	,,	,,	1	Long. ... 17 25 E.
,, ,,	,,	9 17 21	+3 45 57	Chron. ☉	82 43 23	12 35	1 9	−0 40	3	Lat. ... 12 35 S.
										Long. ... 17 25 E.

TRANSIT OF MERCURY ACROSS THE SUN ON THE 6TH OF MAY, 1878.

Date.	Place of Observation.	Latitude.	Longitude.	Time by Chron. to Time of Locality.	Height of Sun. Error of sext $-1'\,25''$.	Later time of Greenwich.	Time of 1st Inner Contact.	Longitude.
		° ′ ″	° ′ ″	Mean of 4 H. M. S.	Mean of 4 ° ′ ″	H. M. S.	In the Chron. H. M. S.	° ′ ″
6 May, 1878.	Belmonte	12 22 40	16 49 24	10 6 50	74 36 55	3 39 39	11 35 29	16 50 15

It is worthy of note that the first longitude I determined in Belmonte by the chronometer was very near the true one obtained by the transit of Mercury. This longitude differed likewise but very little from that obtained by the eclipse of the 1st satellite of Jupiter on the 23rd of April.

I do not include in this table the numerous observations made to check the movements of the chronometers, which I propose some day publishing separately.

The great difference observable between the readings of different chronometers proves that it must have arisen from the chronometers themselves.

As will be remarked, the instrument used by me was the sextant with the artificial horizon of mercury; I had no other, as the aba, the only universal theodolite brought out from Europe, remained in possession of my companions.

My sextants came, one from Casella of London, reckoning 5″, and the other from Lorieux of Paris, reckoning 30″. My azimuth compasses were manufactured in Berlin, and had belonged to the unfortunate Baron de Barth.

My chronometers were made by Dent of London, two being algebraic, and another, marine, which was sent me to the Bihé from Benguella.

The last was an inferior one: but the other two were excellent, more especially that which I distinguish in the calculations by the letter S.

Many of the altitudes were fixed by the hypsometer, and others by the aneroid, compared with the hypsometer.

The altitudes are marked upon the map in English feet.

The Map of the Bihé country, undoubtedly very rough and incomplete, was drawn up by the compass during my sporting excursions; but even as it stands it possesses sufficient correctness to enable an opinion to be formed of the territory.

I suspend for the moment any further mention of the details of my maps, in order to give a rapid sketch of the country they represent.

From Benguella to the Dombe, as will be seen, I followed the coast over calcareous ground which abounded in various mineral ores.

The tract is badly supplied with water during the dry season, and the valley of the Dombe Grande has scarcely sufficient to render it greatly productive. Vegetation there without being actually poor, does not possess that richness which is peculiar to the intertropical countries. Between Benguella and the Dombe the only drinking water obtainable is from a small marsh in the Quipupa.

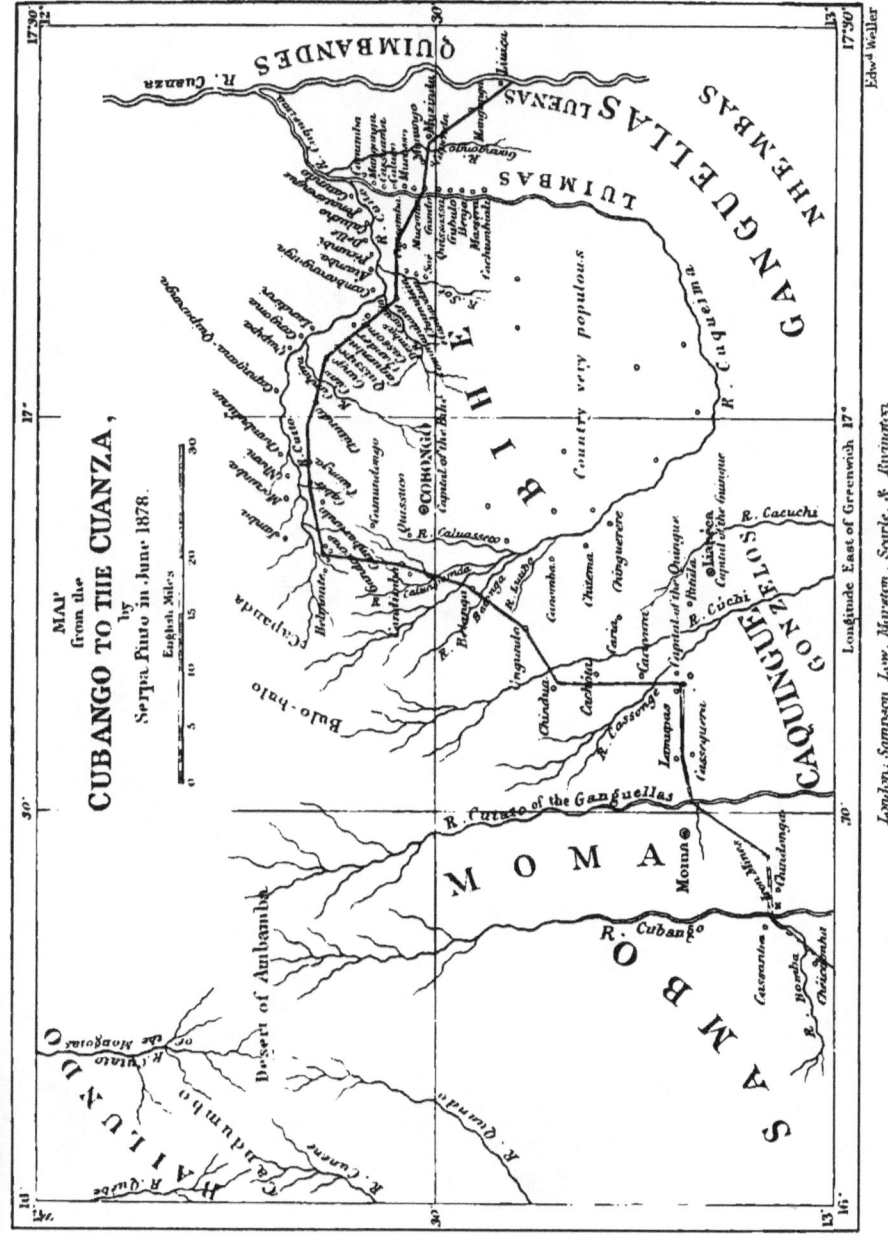

The territory is abundant in game, and a great variety of antelopes is observable, the most common being the *Strepsiceros Kudu*, the *Cephalobus mergens*, the *Cervicapra bohor*, and the *Oreas canna*. The rocks of carbonate of lime which form the orographic system of the Dombe Grande abound in *hyrax*, and among the large and splendid plantations of manioc which cover the plain, great numbers of *hystrix* find a dwelling; they are somewhat larger than the European ones and cause great devastation to the cultivated grounds. The valley of the Dombe Grande is certainly the best portion of land in the province of Angola. On the score of salubrity there is not much to complain of, and the soil is of great fertility. A seaport, the Cuio, is only a few miles from the best centre of production.

The mountains which frame the valley are full of minerals and have in many places been worked, but always on a small scale owing to the want of capital. Both sulphur and copper are to be met with there.

The native population are well-disposed and laborious, at least in so far as the blacks left to their own devices are ever likely to be.

Between the Dombe and Quillengues the country is deserted. By the road we took, water was wanting; but vegetation, which was poor at the outset, assumed a most luxurious aspect the nearer we approached Quillengues.

By following the course of the river Coporolo, there is no deficiency of water, and I heard it stated that a rich vegetation extends all the way; and yet the country in those parts is uninhabited.

On leaving the Dombe the land rises suddenly to a height of 1804 feet, and a system of mountains commences, running north and south with little valleys in between. These mountains continue gradually rising, till near Quillengues their summits attain an elevation of 2950 feet. It is in the river Canga that the granitic formation begins, and with it a more abundant vegetation. All the rivers marked upon the map up to Quillengues are little better than mountain torrents in the rainy season; still, in many of them, it is possible to find water in the summer by simply digging holes in their sandy beds. The Coporolo itself is liable to this condition of poverty.

Quillengues is an extensive and fertile valley, very similar in character to the Dombe, but naturally of infinitely less value, owing to want of communication with the coast.

Its population is dense, and on its meadow lands thousands of heads of cattle, of excellent breed, find abundant pasturage.

The Quillengues are a robust and warlike people, and in the attacks they make upon the Mundombes, they always come out victorious, which does not, however, prevent them being vanquished in turn by the people of the Nano country, who make descents upon their territory and carry off cattle and prisoners.

The Quillengues as well as the Dombe tribes are subject to the King of Portugal, but are not so submissive as the Mundombes.

There is little doubt but that the country of Quillengues has a prosperous future before it, attainable so soon as easy communications are opened with the coast, with Huila and Caconda, and a proper administration governs its affairs.

From Quillengues to Caconda the proper track runs through Caluqueme, a thickly-peopled territory; but I selected another one for motives which I explain in my narrative.

On quitting Quillengues in a S. E. direction the traveller arrives at the lofty mountain chain of the district, where the ground rises rapidly to a height of 5725 feet; the part where I crossed it bears the name of Mount Quiccoua.

There commences the vast and lofty table-land of South Central Africa, and thence to the Bihé the enormous plain maintains the same altitude, with the slightest possible depressions in the beds of the rivers, and a trifling system here and there of isolated mountains.

From this table-land run permanent streams, the first I fell in with being an affluent of the Cunene.

The arboreous vegetation upon this high ground is quite as strong as at Quillengues, but the herbaceous is richer, if it is possible to be so.

The ground continues granitic and a greater abundance of termites begin to appear. The only villages that are met with on the road are Ngola and Catonga, which I have fully described elsewhere.

In Caconda the country is somewhat more undulated, and should be no less rich and productive than that around Quillengues.

It is cut through by permanent rivers which irrigate it in every direction: these run into the Catapi, an affluent of the Cunene.

Miasmatic fever is endemical in Caconda, precisely as at Quillengues and on the coast, but it exhibits a milder character in the former district and but rarely claims any victims.

As regards climate, Caconda differs essentially from the coast and even from Quillengues.

Though situated at only 13° 44' from the equator the climate, which should be excessively hot, is tempered by the enormous height at which the territory lies; but it is on that account subject to those sudden changes between day and night which are prevalent throughout the table-land. There is a constant struggle going on between the altitude and latitude, the result of which is, that the dominion of the latter is most sensibly felt during the day, when a vertical sun darts down its rays of fire, and the former reigns supreme at night, at an elevation of 1860 feet above the sea-level.

I remember that Anchieta laughingly observed that one might be perfectly comfortable in Caconda if an apparatus, in contact with a thermometer, could be invented to put additional blankets upon one's bed, while asleep, in proportion as the thermometer descended.

This vast disparity of temperature between the day and night becomes most sensible when the sun has a northern declination; as when the orb is going towards the south of the equator it is much diminished.

I was informed over and over again that all the fruits of Europe are produced in Caconda, but unfortunately I cannot state the fact of my own knowledge as I did not fall in with them; still I have reason to believe that they might be acclimatised. The potatoes are very good and abundant, not only in Caconda, but throughout the table-land; their transport to Benguella is, however, so difficult under present arrangements, that all the potatoes consumed there are brought from Lisbon.

European potherbs and vegetables are plentiful and good.

Near the fortress the population is scanty, but at a certain distance it is dense enough, and is there governed by independent chiefs.

From Caconda to the Bihé the country is very thickly peopled, and if fewer cattle are raised than on the other side of Caconda, agriculture is somewhat more attended to.

In the countries of the Nano, Huambo, Sambo, and Moma, the natives are savage, warlike, and independent.

The land, as will be noticeable by the map, is intersected by various streams, the waters being drained into three great arteries, the Cunene, Cubango, and Cuanza.

To the north of the Sambo territory there is a vast tract of waste ground, called in the country the *Enhuna* de Ambamba; this is for the most part marshy, and is the source of five important streams, two of which run northwards, and three towards the south.

Of those which flow in a northerly direction, one is the Quebe, which debouches into the sea at 10° 50' S. latitude, near the Tres Pontas, between Novo Redondo and Benguella Velha.

This river assumes at the lower part of its course the name of Cuvo. The other is the Cutato dos Mongoias, which flows northwards and becomes an affluent of the Cuanza.

The three that run to the south are the Cunene, the Cubango, and the Cutato dos Ganguellas, which unites with the Cubango.

The most important system of mountains which I met with, was the chain running from N.E. to S. W., to the north of the country of the Huambo, from whose slopes spring the Calac and the Cucuce, which unite and then flow into the Cunene.

A rough observation made with the aneroid proved the summit to be upwards of 8200 feet above the level of the sea.

By way of exception to my rule not to baptise any rivers or mountains in Africa, I bestowed upon this chain the name of Andrade Corvo, it being described in the country merely as the Huambo chain.

I could not discover among the natives any traces of the existence of other ore than that of iron, by which I do not desire it to be inferred that there is no other.

The soil is still granitic, and all that can further be said of it is, that in many places it is of animal formation, being produced by the labours of termites.

Besides the special disposition which I met with, and remarked upon, in the termitic ground on the banks of the Cutato dos Ganguellas, there are four distinct termitic formations, which I presume to belong to four different species.

The slopes of the Andrade Corvo chain, between the Calac and the Cucuce, abound with game; indeed, with the exception of the Zambesi, I never saw a greater quantity in Africa.

In addition to the antelopes which I before referred to, when speaking of the Dombe, there are abundance of the *Hippotragus equinus*, *Catoblepas taurina* and *Bubalus Caffer*.

The forests are in great part formed of dicotyledonous trees, with innumerable species of acacia. Of creeping or climbing plants there are extremely few.

Crossing the divisional line of the waters between the Cubango and the Cuanza we enter the Bihé country, certainly the most important in the south-west part of Central Africa.

This country, whose people I described at length in the foregoing chapter, is intersected by two important rivers, although they are unnavigable, viz. the Cuqueima and the Cuito. Innumerable rivulets water the land in all directions, and form affluents of these main arteries.

The climate is similar to that of Caconda, and the same atmospheric conditions are observable in both places.

The soil is granitic, and of wonderfully productive power. The pasturage is excellent for sheep and cattle; the country is poor in game; but by way of compensation there are few or no wild beasts.

I do not think it is rich in mineralogical products, inasmuch as, notwithstanding the density of its population, no vestiges of any rich mineral ores appear among the people; which would scarcely be the case if they existed, as I have ever observed in Africa that the first to discover gold, copper, lead or iron, were the natives.

Fig. 29.—Ant-Hills, found between the Coast and the Bihé. 1 and 2 are a few inches above the ground. 3 and 4 are from 3 to 7 feet high.

What is really rich in the Bihé is the soil, and I know of no African country more susceptible of prosperity through agriculture and trade than that territory.

The European race could reside there in the utmost comfort, and the offspring of such as have settled in the country, and become connected with the natives, is physically admirable.

During my stay in Belmonte, I made a careful study of its climatological conditions; and more especially during the first month when that pertinacious rheumatism contracted during my journey, prevented my quitting the house, I regularly observed the barometer and thermometer every three hours during the day.

I exhibit below a Table of those observations over a space of thirty days, and at the same time remark that the equableness of temperature notable during the day is owing to the season of the year in which the observations were made, a season which corresponds to our autumn.

The rains fall at two different periods, with an interval of fine weather between them, occurring in December and January. The first rains commence about the middle of October, and continue till the beginning of December; they are more moderate than the second, which fall from the end of January to the commencement of March.

The prevalent winds are from the east, and at times they blow very persistently and strongly from that quarter; this however is in the dry season, for during the rains, the heaviest storms I remarked came from W.S.W. and from the south. The rains are always, and more especially in February, accompanied by electric meteors, which fall in the midst of terrible thunderstorms.

The Table below exhibits my observations from the 25th of March to the 23rd of April, 1878.

Year 1878.		6 o'clock.		9 o'clock.		Noon.		3 o'clock.		6 o'clock.	
Month.	Day.	Barometer.	Thermometer.	Barometer.	Thermometer.	Barometer.	Thermometer.	Barometer.	Thermometer.	Barometer.	Thermometer.
March	25	629·8	19·1	630·5	20·4	629·2	22·4	628·8	23·2	630·0	21·6
"	26	632·0	20·1	631·9	21·2	630·8	21·6	629·8	21·5	629·5	21·0
"	27	629·5	19·4	632·0	19·9	629·6	21·0	628·5	21·3	630·0	20·6
"	28	630·0	19·4	631·6	19·9	629·5	20·4	629·0	22·1	629·0	21·6
"	29	630·2	20·6	632·3	20·8	630·0	21·6	628·5	22·5	629·2	22·1
"	30	631·0	18·3	632·0	20·6	631·0	21·9	630·0	22·2	629·9	21.3
"	31	631·0	19·2	632·3	20·0	631·2	20·9	629·2	21·3	631·0	20·4
April	1	630·5	18·6	632·0	19·5	630·6	20·4	630·0	19·9	630·0	19·8
"	2	631·0	17·5	632·0	18·7	630·0	21·1	629·3	20·2	630·0	20·2
"	3	630·0	18·8	632·5	20·0	630·5	21·1	630·0	21·2	629·0	20·9
"	4	632·0	18·6	632·0	20·2	630·0	21·2	629·5	21·6	630·0	20.7
"	5	630·0	18·8	632·0	20·0	630·3	21·1	630·0	22·0	629·8	20.1
"	6	630·0	17·2	632·3	19·8	631·0	20·4	630·5	21·7	630·0	20·2
"	7	630·0	17·8	632·0	19·7	630·5	21·0	629·0	22·7	630·0	21·5
"	8	629·0	17·6	632·0	19·9	630·0	21·5	629·5	22·8	630·0	21·3
"	9	629·5	18·4	631·5	20·4	631·0	21·8	629·3	22·6	629·8	21·1
"	10	631·2	18·1	632·8	20·5	631·5	21·7	629·4	22·4	630·0	21·5
"	11	630·5	16·6	631·9	20·2	631·0	21·4	629·5	23·0	629·8	21·7
"	12	629·0	16·4	629·9	20·1	629·0	21·1	627·0	22·6	629·0	21·8
"	13	628·3	18·2	630·0	20·2	629·6	21·6	629·4	22·3	629·5	21·1
"	14	629·0	18·6	631·5	20·4	630·6	22·0	629·5	23·1	630·0	21.7
"	15	631·4	17·2	632·6	19·7	631·0	21·3	630·5	22·4	630·5	20.7
"	16	630·6	16·1	632·0	19·0	630·3	21·3	629·0	22·8	630·0	20·2
"	17	632·6	19·4	633·0	20·7	631·0	22·0	630·0	22·2	630·0	20·0
"	18	631·6	18·0	632·0	20·1	630·0	21·0	629·7	22·7	629·9	19·8
"	19	631·2	17·8	632·2	20·3	630·6	21·0	630·1	23·0	630·5	19·7
"	20	630·7	16·5	631·9	20·1	630·4	21·2	630·0	22·7	630·0	20·1
"	21	631·0	15·6	632·1	17·8	630·3	19·8	629·3	20·6	629·8	19·5
"	22	630·0	14·6	632·0	17·1	630·0	19·2	628·7	20·4	629·0	19·4
"	23	630·3	14·9	632·0	17·9	630·5	20·0	629·2	21·3	630·0	20·0

By this series of observations, it will readily be seen how mild is the climate of the Bihé during this period of the year.

The diurnal advance of the barometer is very remarkable, remaining unchangeable in presence of the sudden changes in the atmosphere.

A meteorological table drawn up at 0h. 43m. of Greenwich, or 1h. 50m. of the Bihé, will complete the atmospheric study of this country during the period under review.

Meteorological Table made at 0h. 43m. of Greenwich or 1h. 50m. of the Bihé.

Month.	Day.	Barometer.	Thermometer dry.	Thermometer wet.	Rain in millimetres.	Direction of Wind.	State of the Atmosphere.
March	25	628·7	22·9	20·2	40	S.S.W. weak.	Thunder during the night; to-day sky clear.
,,	26	629·6	22·1	20·0	2	W.S.W. weak	Cloudy at night; fleecy clouds by day.
,,	27	629·1	21·0	20·1	31	E. strong	Rain during the night.
,,	28	628·8	21·5	21·2	0	Calm	Some clouds and fleece.
,,	29	629·0	22·3	21·6	0	,,	,, ,, ,,
,,	30	630·0	22·0	21·0	0	,,	,, ,, ,,
,,	31	629·5	21·5	20·8	0	E. strong	Cloudy.
April	1	630·5	20·2	19·4	17	Calm	Cloudy. Thunder at night from the N.W.
,,	2	629·3	19·8	19·1	0	E. strong	Some clouds and fleece.
,,	3	630·0	20·9	19·1	0	E. moderate	,, ,, ,,
,,	4	630·3	21·5	20·2	0	,,	,, ,, ,,
,,	5	630·5	21·8	20·6	0	,,	,, ,, ,,
,,	6	630·0	21·1	19·2	0	,,	,, ,, ,,
,,	7	629·3	21·8	19·7	0	,,	,, ,, ,,
,,	8	628·1	22·5	19·8	0	,,	,, ,, ,,
,,	9	629·6	22·2	20·6	0	Calm	,, ,, ,,
,,	10	629·0	21·8	19·9	0	,,	Sky clear.
,,	11	629·8	21·0	19·8	0	,,	,, ,,
,,	12	627·8	21·8	19·8	0	,,	Some fleecy clouds.
,,	13	629·5	22·0	20·1	0	,,	Cloudy.
,,	14	630·0	22·5	20·2	0	,,	Some fleecy clouds.
,,	15	630·5	21·6	19·6	0	E. strong	Sky clear.
,,	16	629·8	21·6	19·7	0	Calm	Some fleecy clouds.
,,	17	630·0	22·0	18·6	0	E. strong	,, ,, ,,
,,	18	630·0	22·2	20·3	0	,,	,, ,, ,,
,,	19	630·4	22·5	20·1	0	E. moderate	,, ,, ,,
,,	20	630·2	22·0	20·2	0	,,	,, ,, ,,
,,	21	629·8	19·9	15·5	0	,,	Sky clear.
,,	22	629·6	19·9	16·1	0	,,	,, ,,
,,	23	630·0	20·5	18·3	0	E. strong	,, ,,

The foregoing figures show the results of thirty days' observations, and I continued the same system during the entire journey, saving only where interrupted by illness or occasional disturbances.

The land from Belmonte eastward slopes a little down to the Cuqueima at that part where the river runs from south to north. On the right bank of the Cuqueima it deviates somewhat in its descent to the valley of the Cuanza.

In the eastern part of the country, the arboreous vegetation reappears in all its wealth of foliage, and there are small but dense forests visible here and there.

Throughout the vast territory comprised between the Bihé and Benguella the tsee-tsee or titsee fly, that scourge of so many parts of South Africa, which by destroying the horse and ox deprives man of two of his best auxiliaries in practical life, is entirely unknown.

A species of epidemic, called in the country *cahonha*, attacks both cattle and sheep, but without committing anything like the ravage arising from similar causes, noticeable in Europe and other parts of Africa.

The *Horse sickness*, which kills so many animals in the Transvaal and in the Calaári, does not exist in this territory. Swine seem to prosper quite as well as in Europe, and the people are able to preserve the meat without difficulty, which they cannot do near the sea.

The country as far as the Cuanza, and even beyond that river, is entirely without salt, all that is used there being brought from the coast.

There are no mines of rock-salt, and the waters, including those of the lakes, are all drinkable.

In the foregoing brief summary, I have endeavoured to condense the results of my observations, and give a general idea of the country; I will now conclude with my brief opinion concerning it.

Placed in a geographical position very different to that of the Transvaal, the tract of territory comprised between the coast and the Bihé approximates thereto in the way of climate, and possesses a more fertile soil. A comparison between the same plant growing in the two countries makes this very evident.

It has a native population far more condensed than that of the Transvaal, and infinitely more agricultural. It is not less abundant in good pasturage, and is richer in woods and forests.

The Transvaal, it is true, possesses great mineral wealth, which is wanting here, but I am of opinion, notwithstanding, that a more prosperous future is in store for this country than for the Transvaal, inasmuch as the latter is isolated from the rest of Africa by arid deserts and the tsee-tsee fly, while the former is in easy communication with the other territories of the interior, whose natural wealth is perhaps greater than its own.

CHAPTER VII.

AMONG THE GANGUELLAS.

Passage of the Cuanza—The Quimbandes—The Sova Mavanda—The Rivers Varea and Onda—Tree-ferns—Tribulations—Slaves—The River Cuito—The Luchazes—Emigration of Quibocos—Cambuta—The Cuando—Leopards—The Ambuellas—The Sova Moem-Cahenda—Descent of the River Cubangui—The Quichobos—Sudden Changes—I start for the Cuchibi.

On the 14th of June, as I had determined, I broke up my camp, and at ten o'clock commenced the passage of the Cuanza, which took a couple of hours.

Fig. 30.—Crossing the Cuanza.

My mackintosh boat, purchased in London, did me

the greatest service; and I had also four canoes which were lent me by the Sova of Liuica.

The passage was effected without the slightest accident, and by noon I was able to continue my journey which I did in an easterly direction, penetrating into the country of the Quimbandes. Having passed near the villages of Muzeu and Caiaio, I encamped at about two hours' journey E. S. E. of the village near the source of the Mutanga rivulet, which runs N. W. into the Cuanza. I noticed that the villages in these parts were

Fig. 31.—QUIMBANDE MAN AND WOMAN.

not nearly so strongly fortified as those on the other side of the Cuanza. The Quimbandes form a confederation, their country being divided into small states which always combine for the common protection. The whole of the numerous villages around my camp were under the sway of the Sova Mavanda, who is himself a tributary of the Sova of Cuio or Mucuzo, situated on the banks of the Cuanza but more to the northward. The sight which first struck my attention among the Quimbandes was the head-dresses of the women, the most extraordinary I ever beheld in my life. Some arrange the hair in such a way that—after it is embellished with cowries—it looks for all the world like an European woman's bonnet. Others friz it out, and twist and turn it, till it wears the aspect of a Roman helmet.

Cowries seem to be profusely lavished in the adornment of the female head, and white or red coral is also visible, but not to the extent observable among the people to the west of the Cuanza.

The hair in these stupendous head-dresses is fixed with a most nauseous red cosmetic, formed of a resinous substance reduced to powder and castor-oil.

Castor-oil is prepared in great quantities among these

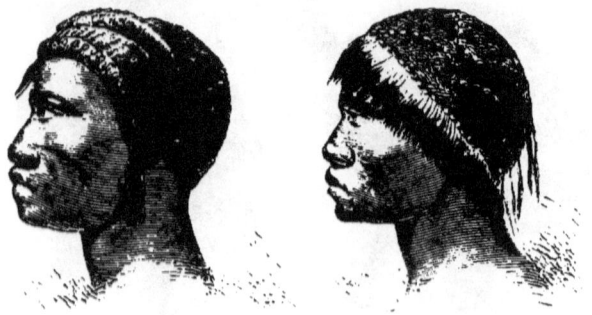

Fig. 32.—QUIMBANDE GIRLS.

people. After extracting the seeds of the *Ricinus communis*, they dry them and then reduce them to powder. This powder, kept for several hours in boiling water, furnishes the oil, which, when cold, is roughly separated from the water and preserved in small calabashes.

The oil is not used by the natives as a purgative.

I speedily remarked that the feminine type among the Quimbandes approaches somewhat to the Caucasian, and I saw some women who would have been called pretty if they had not been black.

Immediately upon my arrival I sent a small present

to the Sova Mavanda, who was profuse in his thanks, although he pressed me further to give him a shirt.

A like request had already been made me by others, proving a tendency in the direction of body-covering.

The male natives cover their nakedness with two aprons of small antelope skins, which they suspend before and behind from a broad belt of ox-hide. The Sovas alone use leopard skins. As to the women, they go almost naked, and a fragment of cloth does duty for the traditional fig-leaf of our mother Eve.

Early on the following morning some of the chief's porters came to inform me that the men I was expecting arrived the night before on the other side of the Cuanza, where they were encamped.

I did not give the slightest credence to the news, knowing so well, as I did, the habit of the people to tell you anything they think may be agreeable to you in order to extract a reward for the intelligence. Still I told the messengers, who averred they had seen the men, that I would suitably reward them if they would procure me some token from Dr. Chacaiombe assuring me he was on the track.

That same morning the Sova Mavanda sent certain envoys to inform me that he was going to set out immediately to attack a neighbouring village, where one of his subjects had revolted against his authority, and to beg me, at the same time, to aid him in his campaign. I of course refused to render him any assistance, but I did so in a mode that prevented him feeling angered at my neutrality.

It was about mid-day when Mavanda's army passed near my camp.

In front was carried a tricoloured flag, like that of France but with the colours reversed, fluttering from a lofty staff. Then came two men carrying an enormous powder chest, by means of a rope and pole. By the way they shouldered it, it was evidently empty. They were followed by the Sova, surrounded by his grandees or staff, and after them came the army in single file. There might have been some 600 men armed with bows and arrows, and eight carrying muskets. A few steps ahead of the flag were a couple of blacks beating war drums and making the most horrible noise of which they were capable.

At nightfall the army returned without having had an engagement, as the enemy had surrendered at discretion.

On reaching my camp they did me the honour to treat me to a sham fight.

The bowmen spread out in one long line having the flag in the centre, and behind it the powder chest and the Sova.

This single line, for each man was isolated, gradually began to surround the imaginary village they were attacking, and contracted as it grew nearer.

Then, at a given signal by their chief, the soldiers rushed upon the village, running and bounding in the air, and uttering at the same time the most frightful cries to intimidate their adversaries.

When I thought they were going straight to their homes to attack their supper, I observed them return to the position they occupied before the fight, and again collected at the command of their sovereign, they re-entered the village in the same order in which they had left it.

Shortly after this exhibition was over the messengers who had previously called upon me came again to say they had seen my doctor, but that he would not give them any token for me. This convinced me that my suspicions were right, and that there was not a word of truth in what they had told me.

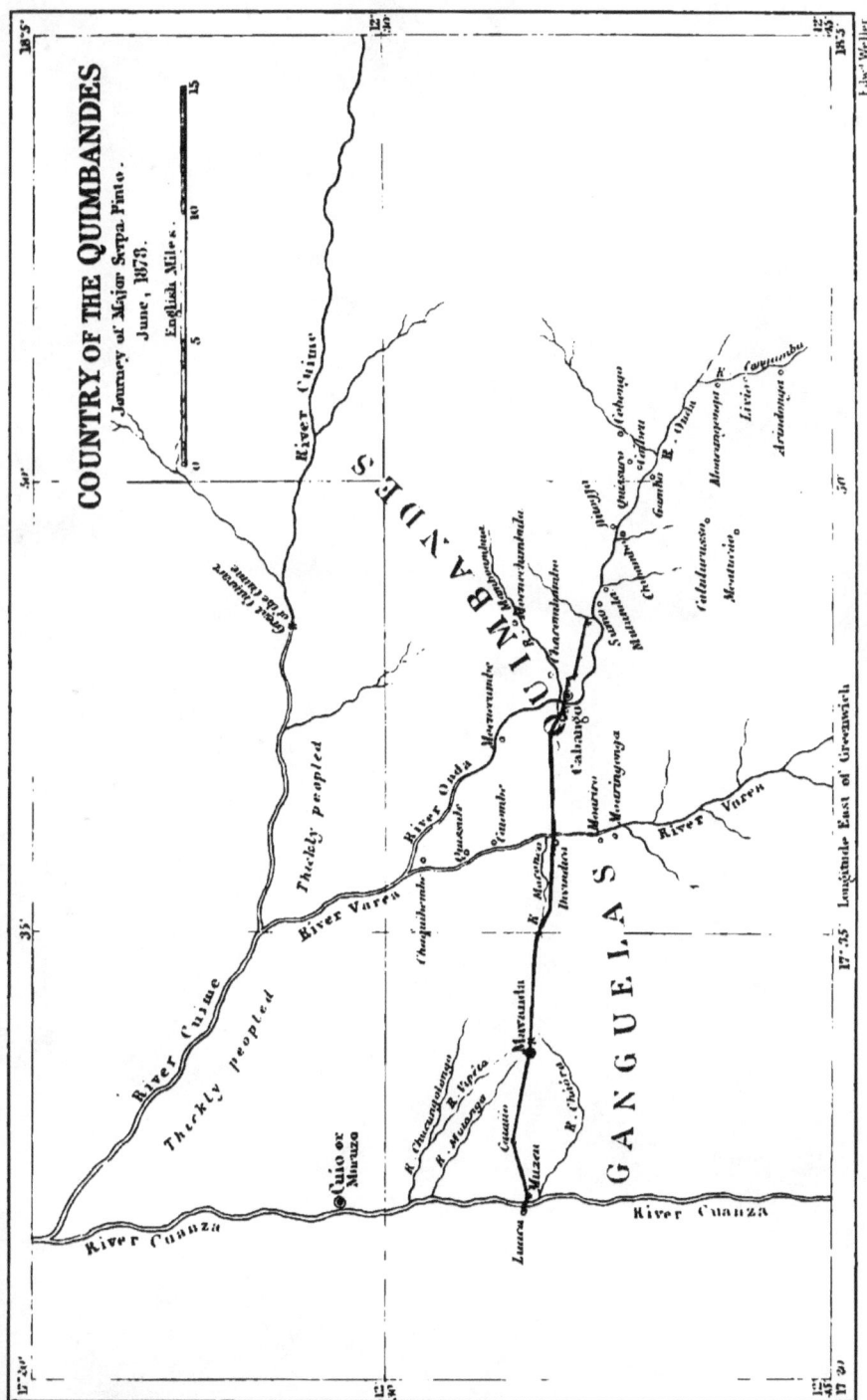

My encampment began to cause me serious alarm, for being covered with dry grass it might catch fire at any moment; and as my blacks were often shivering with cold they did not calculate the danger, but kept up enormous fires in the huts.

Between the river Cuqueima and Mavanda and even beyond, the sugar-cane and cotton plant grow vigorously. The Quimbandes cultivate cotton, which they spin into threads on which to string cowries and beads.

On the following day the natives again asserted that the carriers were on the banks of the Cuanza and could not pass the river for want of canoes.

I resolved to send Augusto thither accompanied by a Quimbande guide.

At eleven o'clock an envoy arrived from the Sova to announce a visit from the latter.

Shortly after, Mavanda arrived surrounded by his court; and if he evinced surprise at sight of me, I am sure I must have done so at sight of him, as he was certainly the biggest man I had ever beheld in my life. To an enormous height he added a trunk of truly phenomenal proportions, and was otherwise inordinately fat. Round his huge waist was twisted an old cloth, from which hung three leopard skins.

Several amulets were dangling from a collar of beads round his massive throat.

It would seem as if Mavanda, being big himself, delighted in things upon a large scale, for he made me a present of the largest ox I ever saw in Africa.

After the customary compliments he said quite abruptly that he had come to ask me a favour, which was, to give him a "remedy" to save his herds of cattle. These animals, being sent out to pasture, strayed away, and not all of them returned to their shelter at night: but, wandering in the woods, were devoured by wild beasts or otherwise disappeared.

I furnished him at once with a remedy in the shape of a piece of advice, viz. to employ herdsmen who, instead of allowing the cattle to stray wheresoever they chose, might lead them to their pasture, and bring them home again in the evening. This was a new idea which struck him as not a bad one, and he said that, although it was contrary to the customs of the country to watch the herds, still it should be done, and he would begin at once so as to save his beasts.

Fig. 33.—THE BIHENOS CONSTRUCTING HUTS IN THE ENCAMPMENTS.

I showed him a barrel-organ, exhibited my rifles, fired them off before him and observed with amusement the surprise and wonder depicted on those huge but good-natured features. He retired in the evening, and we parted mutually pleased with each other.

He was no sooner gone than envoys arrived from the Sova Capoco with a letter for me. It furnished me with news about Chacaiombe, told me he had sent carriers, and begged me to allow one of his caravans, that he was desirous of despatching for purposes of trade to the Zambesi, to travel with me.

This letter decided me upon remaining in my present quarters for some six or seven days to wait for the carriers, although I did not even now lay very much stress upon their arrival, and it was in this sense that I answered Capoco's epistle.

My resolution being taken, I ordered my encampment to be reconstructed, and the huts to be covered with green boughs as a protection against fire.

Fig. 31.—Skeleton of a Hut.

The following morning, therefore, there was great activity in and about the camp, which by noon began to assume quite a pretty aspect.

It was composed of conical huts made of the trunks of trees, each hut measuring ten feet in diameter at the base by eight feet high.

My own hut, built by the Bihenos with more care than the others, measured sixteen feet in diameter and was twelve feet in height.

The encampment was formed by a circular line of huts, connected by a hedge of thorny trees.

My dwelling occupied the centre, and in front of it were piled the goods. My immediate attendants arranged their huts all round me and within call.

The labour of constructing the camp was just at an end when I was informed that some messengers of the Sova of the Gando wanted to see me. I ordered them to be shown in, when I immediately recognised in one of them a grandee of that chief whom I had seen by his side when passing through the country. They brought me a letter and a parcel which some petty chief or other had forwarded to the Sova for me.

Fig. 35.—HUT BUILT IN AN HOUR.

I opened the letter and found it was from my friend Galvão da Catumbella, accompanied by a little present, which he had addressed to the Bihé under the impression that I was still there.

The cordial feeling which I had managed to awaken among the people I had passed through, had produced a good result, and the letter and packet thus reached me in safety, passing from hand to hand.

I opened the parcel and found among other things a box of Malaga raisins, a very welcome gift, as they helped to relieve a little the monotony of my already very poor provisions.

The letter gave me some European news, the last I obtained until arriving at Pretoria. The sight of the lines awakened in me the thought, accompanied by a great sinking of the heart, of how long and weary a period must yet elapse ere I could receive intelligence about those who were so dear to me, and I laid my head on my pillow that night with uncomfortable presentiments in respect of their safety.

At daybreak I had notice given me that a small caravan, commanded by a black, and carrying wax from the interior, was passing on its way to the Bihé. I sent for the head-man and requested him to convey a letter for me to that place, there to deliver it to someone who could forward it to Benguella. He agreed, but asked me to be quick as he intended to sleep that night beside the Cuqueima.

My time was short, and who should I write to? I must not lose the chance offered me by this unexpected messenger of assuring my loved ones that I was still alive.

I seized my pen and traced a few hasty lines to Doctor Bocage. I enclosed therein two small notes, one to my wife, the other to Luciano Cordeiro.

The leader of the caravan, who had already become impatient at the delay, received the letter and departed.

I now know that the packet reached Europe in safety and was received by the person to whom it was addressed, but I never learned how it was forwarded from the Bihé to Benguella. Doubtless it owed its safety to the protection and friendship with which I was honoured by Silva Porto.

The Sova Mavanda passed the day with me, and we had a long talk. I gave him various little articles, and among others a box of lucifer matches, with which he was both astonished and delighted.

When he retired he said to his *macotas*, in a tone and in words which I have not ceased to remember:

"You see afar off a bird which soars aloft and then alights upon a distant tree, and you say it is a dove, then you walk on until you are quite near, and are astonished at its size, for it is an eagle. Thus it was with the Manjoro," (a name they bestowed on me) "when far from our village, we said he was a dove; now we live with him and know him, we find he is an eagle!"

During my rambles in the neighbourhood, pursuing the antelopes, which were scarce, I drew up the map of the country, or rather, was able to complete the map of the territory lying between the Cuqueima and the Cuanza.

The Sova Mavanda sent to inform me that the greatest favour I could do him was to give him a pair of trousers. I resolved to humour him, but having nothing that could fit those stupendous limbs within many ells, I called in old Antonio, and much to his astonishment turned him into a tailor and sent him to measure his Majesty for the wished-for garment. I then cut out the pantaloons and set Antonio to work to stitch them. I cannot say they were a wonderful fit; but they ought to have been big enough, as they took five yards of wide calico! The man was a veritable hippopotamus, though I must say a very good-tempered one.

On the morning of the 20th, an envoy from the Sova came to inform me that as it was the time when the people kept high festival (a species of carnival), his Majesty, to do me honour, would come to my camp, masked, and dance before me.

At eight o'clock some of his attendants arrived and a great concourse of people soon assembled.

Half an hour later the Sova himself appeared, his head thrust into a huge gourd painted white and black, and his enormous body made still larger by an osier frame covered with grass-cloth, likewise painted black and white.

Map No. 3.

A sort of coat, made of horsehair and the tails of animals, completed his grotesque attire.

Immediately upon his arrival the men formed themselves into a line with the attendants behind, and the women and girls removed to a distance. The attendants and men, with upright and motionless bodies, then began a monotonous chant which they accompanied by clapping their hands.

His Majesty took up his station about thirty paces in front of the line and began an extraordinary performance, wherein he acted the part of a wild beast torn with rage, and jumped and capered about amidst the utmost applause from his own people and mine. This lasted half an hour, at the end of which time he ran off at full speed, followed by his men. He reappeared shortly after and returned to my camp, in his ordinary attire, and passed the rest of the day with me. Decidedly I had succeeded in winning his good graces.

I had utilised all the time I could spare from my labours by rearranging the baggage so as to diminish, if possible, the number of the loads. The goods I possessed were of the most moderate quantity and my entire monetary riches consisted of a sack of cowries and of the beads I had purchased of José Alves, but the cost of maintaining my people was great, and I saw, with no little alarm, the diminution of my little store. Game was scarce in the country and small, and with the exception of a few gazelles (*Cervicapra bohor*), it was of little or no account.

How often did not the miserable little heap of goods and beads awaken in me the direst anxiety!

How often did not a shudder of pain and alarm run

Fig. 37.— QUIMBANDE WOMAN CARRYING HER LOAD.

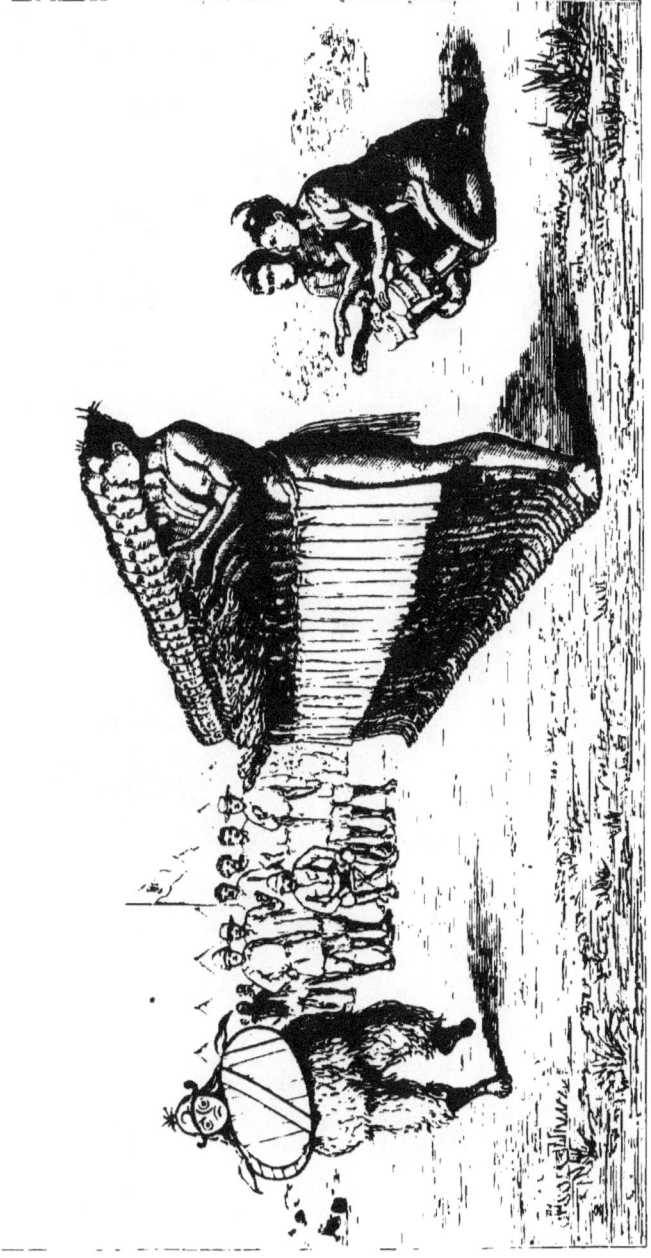

Fig. 36.—THE SOVA MAVANDA, MASKED, AND DANCING IN MY CAMP.

through me as the gloom hanging over my future fate lowered upon my brain!

How often were not the affectionate caresses of my pet-goat Cora, and the chattering attentions of my tame parrot, who flew upon my shoulder to obtain a kindly word, left unheeded!

And yet, just as often, a boundless faith in the work I had in hand would fall upon my aching heart like a balm and banish for the time all anxiety from my mind.

Cold reason would occasionally step in and treat as baseless the rays of hope which were so warm and flattering; but they would not be extinguished, let reason argue as it would, but burned the brighter for the attempt to destroy them.

They are indescribable moments, these struggles of the mind, in the man who stands thus isolated; who is himself the pro and contra of his own ideas, who has no friend by him with whom to exchange his thoughts or from whom to obtain a word of sympathy in suffering and sorrow.

In my youth I had my loves and hates, and with them the pains and heart-burnings that follow in their train—I was a father and saw a daughter I adored pass away from my encircling arms; but never in the past did I feel such an utter void, so deep a depression, as often fell upon me during this African journey!

Alone! perfectly alone as I was in the midst of an ignorant and brawling multitude, whose language and modes of thought were unknown to me, I would sometimes brood until fever and sickness fell upon me, if it were not the approach of those frightful visitants which caused the gloom in my mind.

I do not reckon as suffering the hunger, the sickness, the utter discomfort which assail the explorer in regions like these. Man is, and should be superior to their assaults. The real suffering is doubt and uncertainty; the not knowing how he is to cross the abyss which reason tells him is haply yawning in his path. The real suffering is to see a band of devoted followers accompanying him blindly, and under the persuasion that he who leads will conduct them in safety, walking with him perchance to utter destruction. The real suffering is the tremendous responsibility with which his mission has weighted him. If I were not unwill-

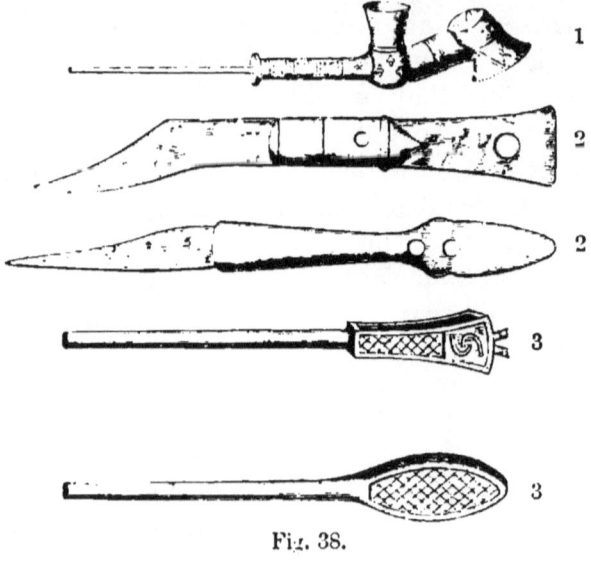

Fig. 38.

1. Pipe. 2, 2. Knives. 3, 3. Tomahawks.

ing that my detractors should experience a little of the pangs of hunger and thirst, of annoyances and privations I had to undergo, I would not have the bitterest of them suffer a thousandth part of the moral torture I myself endured—although it is true that to suffer as I did, a man must not be devoid of sensibility, heart and conscience.

It was under the influence of the feelings I have thus attempted to describe that I traced those lines to

Dr. Bocage, and I greatly doubt whether he considered my letter a cheerful one.

Let me, however, cast aside these reflections, which I fear may not prove very interesting to my readers, and resume the thread of my narrative.

The Quimbandes manufacture various articles of iron and wood in a much more workmanlike manner than the inhabitants west of the Cuanza.

The cold at night was very severe and the difference between the maximums and minimums very great. Notwithstanding the letter I received from the Sova Capôco, I did not give much heed to the promise about carriers or greatly expect the return of my doctor, Chacaiombe. I therefore still continued reducing the loads as much as possible, which could only be done by breaking up some and stuffing their contents into the others. But this, of course, had a limit, the limit of weight the men could carry.

We reached the 22nd of June, the day on which the time I had decided upon waiting for Capôco's carriers expired.

My anxiety was great and I appreciated to the full the troubles dwelt upon by former explorers in having to abandon things which were absolutely necessary, not merely for their comfort but almost for their subsistence.

This question of choice is a very serious one, when every single thing appears indispensable.

As I have before mentioned, I parted long ago with everything intended for my mere convenience, and the few tins of meat which remained I gave to my young negroes as the readiest way of getting rid of them.

My carriers, seeing the difficulty I was in, offered to carry the very maximum weight they could tramp under; but even this unexpected zeal would not suffice to convey the whole of my store, for after all reductions and redistribution of loads, there were four of the latter without porters.

Two of them contained the mackintosh boat, the third a barrel of aguardente, and the last, fifty pounds of gunpowder.

I resolved, with infinite regret, to abandon the boat, and then to ask the Sova Mavanda for a couple of men to convey the powder and spirit from encampment to encampment until two of my porters should have got rid of their loads; which would not be long first, considering the rate at which the stores were being consumed.

The Sova took charge of the boat and gave me the two men I asked for, and all preparations were made to leave next day.

In accordance with this arrangement I broke up my camp on the 23rd at eight o'clock, and after $3\frac{1}{2}$ hours' march arrived on the left bank of the river Varea, which I crossed on a tolerable timber bridge.

The petty chief of Divindica, a hamlet situated on the left bank of the Varea, at the confluence of the Maconco rivulet, put in a claim for toll for crossing the bridge, which I satisfied by the payment of four yards of trade cloth.

The river Varea there runs towards the north, and flows into the Cuime. Its width is 27 yards, and depth 6 feet with but little current, as there are no cataracts near Divindica. I marked as about a mile to the south the villages of Moariro and Moaringonga.

I then travelled eastward, camping at 2 P.M. on the left bank of the river Onda, opposite the large village of Cabango, the capital of the East Quimbandes tribes.

I had with me two bottles of 1815 port, the remainder of a present made me by my friend E. Borges de Castro, and just as we reached the spot where we proposed encamping, the young nigger Mocro, under whose care they were, stumbled and fell, breaking one of the bottles in the act. Imagine my dismay at beholding the precious nectar colouring the ground, without my being able to taste a single drop!

From Mavanda to the sources of the Moconco, whose course I followed until its confluence with the Varea, the trees are perfectly splendid, and the summits of the lofty hills which border the rivulet are very richly wooded. Beyond the Varea the wealth of vegetation is, if possible, even greater.

From the time of leaving the Cuanza I heard the river Cuime spoken of as the largest stream in the Quimbandes territory, an assertion that was confirmed by the important affluents it received, and which raised in me a great desire to inspect it with my own eyes.

From the Cuanza eastward the country presents a very different aspect to that which was observable on the other side. The landscapes are more picturesque and do not exhibit the monotony of the Bihé. The rivers and rivulets dig deeper beds for their waters, and the irregularities of the ground are more sensible. The banks of the streams, whether large or small, appear, beyond the limits of high-water mark, covered with fine trees, while shrubs and underwood form almost impassable barriers in the forests.

In the east part of the Quimbandes territory the population begins to thin. The Sova of Cabango is also a tributary to him of the Cuio or Mucuzo.

The customs of these people are similar to those of the Bihenos, saving in the matter of activity, which among the Quimbandes is changed for the most shameful sloth. They go almost naked, do no work, undertake no journeys, and carry on no trade.

There are but few fire-arms among them, as they have no means with which to purchase them. They gather a little wax, which the Bailundos take from them in exchange for cowries and beads, but the barter is effected upon the smallest possible scale.

The ground is cultivated by the women, and its production is rich. I remarked that manioc and *gingerba* were most prevalent in the plantations.

This country ought to receive particular attention. Bordered by navigable rivers which flow into a large navigable tract of the Cuanza, with a magnificent climate and most fertile soil, where cotton and the sugar-cane, cereals and grass for pasturage, flourish in equal abundance and perfection, inhabited by a people easily subjected, it is in the very best condition for rapid development.

On the 24th of June I crossed the river Onda and encamped on the right bank three miles distant from my last resting-place.

The river Onda at Cabango is about 16 yards wide by 16 feet deep, and runs from the east to the N.W. to flow into the Varca.

After determining the position of my encampment, I took a stroll up the river and met with a good deal of game. I found that above Cabango the Onda speedily narrows to 11 yards, but has a depth of 18 feet with a current of 11 yards per minute; a current, be it remarked, which extends to the very bottom, as I discovered, not only by sounding, but by the inclination of the plants growing there; which was easily visible on account of the crystalline clearness of the waters and the fine white sand forming the river bed.

I saw but one kind of fish in this river, to which the natives give the name of *Ditassoa*; it is not unpleasant eating.

While walking on the river banks I perceived at a distance a group of trees which stood out in fine relief from the landscape and which I took to be palms; on a nearer approach, however, I recognised them as most beautiful specimens of the *Fetus arboreos* or tree fern.

The banks of the Onda are cut vertically and there is the same depth of water at the sides as there is in the middle. It is navigable like the others before mentioned, and therefore presents another natural roadway through this superb country.

On returning to camp an agreeable surprise awaited me, for Dr. Chacaiombe was the first person who met me at the entrance.

I was the more pleased to see him, as his disappearance was one of the black clouds which helped to make parts of my journey so gloomy.

Fig. 39.—Ditassoa—Fish of the River Onda.

I have frequently spoken about Dr. Chacaiombe, and never explained to my readers who he was.

He was the diviner who, it may be remembered, predicted such agreeable things in respect of my future fate, when I was temporarily staying in the house of the Captain of the Quingue's son.

Uniting in his own proper person the functions of medicine-man and diviner, he had come, unsolicited, to attach himself to my staff when in the Bihé, and had never left me until he started on the mission to obtain carriers from Capôco, and whence I thought he never intended to return.

After many words of compliment, Chacaiombe informed me that carriers would arrive within a couple of days, so I resolved to wait for them.

My man Augusto then communicated that the Sova of Cabango had been to pay me a visit, and had gone away much annoyed at not finding me.

I at once despatched my pombeiro, Chaquiçonde, to his Majesty with a request for a couple of men to be

Fig. 40.—Tree-Ferns on the Banks of the Onda.

sent to Mavanda to fetch the boat I had left behind, much to my own sorrow and that of my people, who appreciated the services it had rendered us in crossing the Cuqueima and the Cuanza.

This done, I dried myself thoroughly at the fire

(having arrived very wet from the river), as I remembered with a shudder my frightful attack of rheumatism in the Bihé.

The following morning, early, I went out to seek for game, directing my steps northward, where the country was covered with dense forest. After a walk of some eight miles, I fell in with the river Cuime just below its great cataract. I then turned back, and did not reach

Fig. 41.—Cabango Woman's Head-Dress.

the camp before night, when I was regularly fagged out. My sport, however, had been good, and I had seen the river I so ardently desired to behold, so I soon forgot my fatigue. The stream is certainly a very important one, and, if I am to believe the natives, it is navigable from the great cataract to the Cuanza.

Next day I again explored the Onda, and was greatly surprised at the appearance of a hamlet which I descried at a distance on its banks. On a nearer approach I found that what I took for negro habitations were no other than the residences of white ants (*termites*), collected in considerable groups, with conical tops and having all the appearance, seen afar off, of native huts. On getting back to camp I found the Sova of Cabango, who had just arrived with a suite of sixty men and a great many women.

Fig. 12.—Cabango Man.

Though in almost a complete state of nudity of body, they were extraordinarily dressed about the head. The head-dresses were infinite in variety, in fact were true works of art, and have a technology of their own.

The hair worn by the women, frizzed into the shape of a Roman helmet is called *tronda*, and that which falls in braids on each side of the head is styled *cahengue*.

The male head-dresses, on the other hand, are known under the designation of *sanica*.

The Sova offered me an ox, which I returned in a fashion that seemed perfectly to satisfy him.

On that same day the carriers from Capôco arrived; they were but four, it is true, but four were then enough, two being required for the boat and two others to help carry the heavier loads.

Fig. 43.—Cabango Man.

In the evening my negroes and those of the locality had a jollification, which lasted amid great uproar until past ten.

The cold that night was intense, and the thermometer registered at 3.30 A.M. 0° F. The inequality between the maximum and minimum was most extraordinary, and the dryness of the atmosphere extreme, as my meteorological records show.

The Sova paid me another visit and furnished me

with scraps of information about the country. He stated that he did not recognise the sovereignty of the Sova of Cuio or Mucuzo, but considered himself independent.

There is a good deal of wax about the woods, and the Bailundos come to seek it in exchange for cowries and beads. The natives work in iron and make large hatchets, balls and knives, but their war hatchets, arrows and assegais, are obtained from the Luchazes and their spades from the Ganguellas, Nhembas and Gonzellos.

I discovered that the Sova, whose name was Chaquiunde, was rather loose in his principles and did not adhere very strictly to the truth when it suited his purpose to act otherwise. After another long talk with me, he pretended that he was entitled to a variety of things on the ground of another ox he had given me, which was a pure invention. I saw myself under the necessity of desiring him to leave my encampment; when observing my firmness he changed his tone, and sought to excuse his want of faith by alleging that his Macotas had put him up to the ruse with the idea of dividing among themselves whatever could be got out of me.

Fortunately, about this time, the two Quimbandes arrived with the boat, and I made up my mind to start next day, the 28th.

This resolve I carried out, but not so early as I intended, as the thermometer at 6 A.M. was only two degrees above zero and piercingly cold. I therefore broke up my camp at 8, and after nearly three hours' walk in an E.S.E. direction I stopped again on the banks of the river Onda.

Our marches were of necessity short ones on account of my carriers being so heavily laden.

The ground from the river Varea to that point was

covered with a layer of sand, the subsoil being formed of stiff clay varying in colour from a dirty white to ash grey.

Near the bed of the Onda the soil appeared to be composed of a thick layer of mould, resting upon the same subsoil of grey clay. Beside the river I observed a few ant-hills which were cobalt blue in colour.

The open ground was inhabited by a different species of ants (*termites*) to that which was located in the forests. The former constructed hills with rounded tops, exhibiting the appearance of stumps of trees covered with hemispherical cupola, being from thirty-two to forty inches in diameter at the base by about the same in height. In the forests, on the other hand, they are true cones from two to three inches in diameter at the base, and from ten to twelve inches high.

Being very close together, they have a resemblance to a fence of thorns, stuck into the ground.

These forest ants evidently use in the construction of their dwellings the first material which comes to hand, as the mould forming the surface soil of the woods appears to be that selected, and notwithstanding the cement employed in the fabric, the mounds have not that tenacity and durability noticeable in the hills raised by the ants of the open ground. The latter employ the stiffest clay, and the consequence is, their habitations are nearly as hard as stone. In fact, so strong are they that, though the interior is honeycombed like a bee-hive, a Snider bullet will not penetrate deeper into them than four or five inches.

As I before observed, on the banks of the Onda these ants crowd their hills into limited spaces, and they have, at a distance, a remarkable resemblance to Quimbandes villages.

For upwards of an hour after leaving the encampment I strolled along the river upon open ground, but I then came upon a splendid forest, through which ran several brooks, affluents of the Onda.

At times the forest assumed the aspect of one of those extensive English parks where the ground was completely clothed with a soft green turf. I wandered on and on, until at length my steps were arrested as if by magic, while my eyes contemplated with delight one of the most charming prospects they had ever beheld.

Before me lay in perfect repose a lake of crystalline water, whose bed of fine sand was visible at a considerable depth. Enormous trees springing from the borders of the lake formed an appropriate frame, while the rich, deep green of the foliage, reflected to the smallest bough on the placid surface of the water, greatly enhanced the beauty of the landscape. The green turf to which I have alluded ceased only at the water's edge, and hundreds of birds chirped and twittered amid the dense foliage, and at times skimmed rapidly over the lake.

The natives of the country, who are not much given to poetry or sentiment, are nevertheless sensible of the extreme loveliness of this spot, and call the sheet of water—of which they had frequently spoken to me—by the name of Lake Liguri.

All the rivulets in this territory have marshy banks, and I constantly observed in the stagnant water a red deposit which I at first attributed to the presence of iron, but afterwards discovered that it must be an error, as the green tea made with the water gave no evidence, by the formation of tannate of iron, of the presence of that substance. No doubt the red colour is due to an accumulation of infusory animalcula.

I further observed on my way hither from the Bihé, that all those places which have stagnant water abound in leeches, and that these creatures were still more abundant in the little pools collected beside the affluents of the Onda.

The river continued to be between 11 and 13 yards broad with a depth of 13 to 16 feet, without any very sensible current. Its banks contained a large quantity of game.

On the following day I travelled S.E., still upon the right bank of the Onda, for a space of three hours, forcing a passage with some difficulty through a dense forest, and wading with even greater difficulty, on account of the slimy nature of its bed, through the

Fig. 44.—LAKE LIGURI.

Cobongo rivulet—13 feet wide by 3 deep. After some three more hours I got tired of the Onda, and on meeting another little affluent, the Cangombo, kept along its edge for some distance, then crossed it and encamped on the left bank of a third rivulet, the Bitovo.

On the 30th of June I continued my journey eastward on the bank of the Bitovo, traversed some miles of forest, and then reached the valley of the Chiconde, a rivulet whose course I followed till I reached the Cuito, where I camped. I was much moved on falling in with the Chiconde to observe its waters running rapidly towards the river Cuito, for until then I had only met with streams which ran towards the Atlantic; and their waters, whose ripple and rush had so often lulled me to sleep, were, so to speak, a tie which still bound me to my dear country, as they emptied themselves into the same ocean which bathed the shores of my native Portugal. Could those waters only have conveyed the sighs and whispered words that were uttered over them, how many tender messages would they not have carried to my dear ones!

On leaving the Bitovo, that tie which united me to the Western Coast was snapped, and Heaven only knew whether it would again be joined. That very day a year had passed since I bade farewell to my dear old father; and how vividly did I not remember his parting words and the expression of his fears that we were bidding each other an eternal adieu!

My camp was next pitched in the country of the Luchazes, the Quimbandes being left behind me when I quitted the Bitovo.

Several men and women from the village on the right bank of the Cuito came into camp; but they brought nothing with them for sale, and we wanted food. They promised, however, next day to let us have some canary-seed, as it appeared they did not grow either Indian corn or massambala.

They cultivate in their fields, canary-seed (*massango*), a little manioc, beans, castor and cotton, but all upon a small scale; indeed, barely necessary for the consumption of the growers.

They collect a good deal of wax about the forest, from the hives built in the trees where the bees swarm. This wax they barter for dried fish from the Cuanza, which the Quimbandes bring over, as their own river, the Cuito, apparently produces no fish.

The Luchazes are little given to travelling, and rarely leave their villages except to hunt the antelopes for the sake of their skins. Their field work is carried on by both men and women.

The petty chief who governs the sparse hamlets on the borders of the river Cuito is the Moena-Calengo, who pays tribute to another chief, Moena-Mutemba, the

Fig. 45.—A Luchaze of the Banks of the River Cuito.

situation of whose village I could not precisely ascertain.

The Luchazes work in iron and produce all such implements as they require. Iron is to be found within the country.

One thing particularly struck me among these barbarians, viz. the use of tinder to procure fire, by means of a flint and steel. The flints are imported by the Quibôcos or Quiôcos, and exchanged for wax, and the steels are manufactured by themselves out of wrought iron, tempered by cold water into which they are thrown while the metal is red-hot. The tinder is prepared from cotton mixed with the kernel contained in the stone of a fruit called micha, well crushed.

The Luchazes women use baskets of a different kind to those employed by the Quimbandes, and carry them differently, inasmuch as they are suspended from their heads by a broad strip of the bark of a tree, and fall upon their backs. This mode of disposing of their

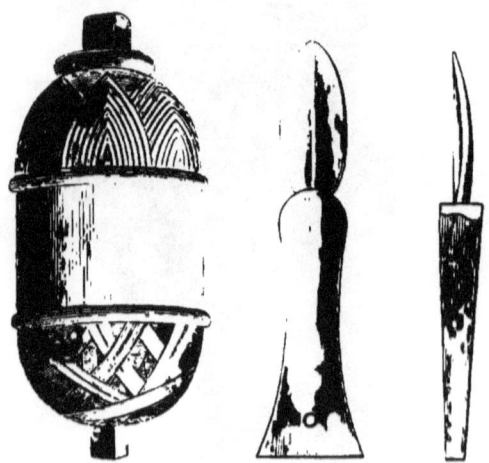

Fig. 46.—Tinder-Box, Flint and Steel.

baskets prevents them carrying their children in the mode generally in use in Africa, upon their shoulders, so that the little ones are slung by their sides.

On the morning of the following day, several women came to offer us some canary-seed (*massango*), but in such small quantities that it rather aggravated than appeased our hunger.

The river Cuito, at the point where I crossed it, is 23 feet wide by 3 deep, and has a current of 82 feet per minute.

It is an affluent of the Cubango, and at its confluence stands the important village of Darico.

It takes its rise in the table-land of Cangala, at no great distance from the sources of the Cuime and Cuiba (affluents of the Cuanza), and that of the Lungo-é-ungo, an affluent of the Zambesi.

Being unable to procure provisions, I resolved to go forward, and while giving orders to break up the

Fig. 47.—A Luchaze Woman on the Road.

encampment, a gang of female slaves, conducted by three negroes, arrived on the banks of the Cuito.

I seized the three blacks, and had the poor creatures set at liberty. When they were assembled in my camp, I informed them that they were free, and that if they chose to join my company, I would find means of sending them on to Benguella.

I assured them they had nothing more to fear from

their guards, and that they were quite at liberty to act as they pleased. To my astonishment, they one and all declared that they did not desire my protection, but wished to continue their course, which I had interrupted.

Whence came they? None could furnish me with an intelligible reply. What then was to be done? I felt a natural repugnance to take them with me against their own will; so, after due deliberation, I resolved to let the poor women follow the sad fate, which they had the means, but lacked the inclination, to escape.

And after all, would they have been better off if they had followed me? It is no easy matter, whatever people may think of it in Europe, to set a gang of slaves at liberty when the operation has to be performed at a distance from European dominion and influence. A batch of slaves consists of natives of different districts, some of which are exceedingly remote. If he who sets them at liberty is desirous of restoring them to their families, he will have to traverse a great part of Africa in search of the homes of his protégés, which is practically impossible.

To abandon them after giving them their freedom is tantamount to delivering them up as a prey to the first tribe which they fall in with.

It frequently happens that these unfortunates, carried off from their villages in tender years, lose all recollection of the place where they were born, and learning to speak a language different to that which they babbled in childhood, end by adopting as their country the land of their slavery, and, in fact, know no other.

Nowadays, that the English and Portuguese vessels of war are cruising in the Atlantic and Indian Oceans, the exportation of human cargoes is considerably impeded; slavery constitutes an object of barter solely in the interior, and the system of the infamous trade is considerably modified.

A slave, for instance, appears in Africa under two guises: either as a prisoner of war, or in payment of a debt due from the parents.

In former times wars were waged for the express purpose of making prisoners and converting them into slaves, nor has the system, unhappily, quite worn out at the present day.

A human creature given by an impecunious parent in payment of a debt contracted or of a fine levied, is common enough.

Where wars occurred in the olden time, every prisoner became a slave, and it was not easy for him, even as an adult, to return to Africa after being once landed on the American coast. The Atlantic Ocean formed a pretty safe barrier.

Those same adults, as being capable of doing much more work, were always preferred to mere youths and children.

But it is not so to-day. A grown man takes to flight, and has ever in his mind the thought of returning to the nest whence he has been dragged, and this hope never abandons him so long as he treads the soil of the continent on which his country lies. As a negro himself said to me:—"*they are always running away.*"

A child, a youth and a woman offer far greater security to the owner, for being more irresolute of spirit, they do not dare face the thought of crossing vast tracts of country to arrive at their own.

It consequently happens that at the present time, in South Central Africa, a child and a woman have greater value, and in the gangs of unhappy creatures who drag their cruel irons across the African soil, it is a rare thing to find a full-grown man.

England, Portugal and France have, of late years, vied with each other in making war upon this trade in human flesh, and the change that time has brought about upon the American continent has aided very considerably in diminishing the horrible traffic and, as a matter of course, in essentially modifying its condition, in Central Africa.

Nevertheless I will venture to assert that it is not this generation or the next that will see the slave disappear from the African soil.

The same principle that was formerly dominant in America of using slaves as colonists exists, and will exist for a long time to come in Africa.

Negro governments have likewise their colonising policy, and we do not possess between them and the places whence the slaves are derived a "silver streak" whereon to float our squadrons and lend these poor creatures the protection of batteries of steel. It is only by the aid of a wide-spread civilisation that we may hope one day to see the end of slavery, but unhappily that day is, as yet, far distant, because the arguments that have been hitherto used have been found far less eloquent and persuasive than rifled shot have proved in the Atlantic and Indian Oceans.

I am myself of opinion that the abolition of slavery in the interior of South Central Africa will only become an accomplished fact when polygamy ceases to exist among the blacks, for although the principles of civilisation may do away with slavery as an institution, the brutal sensuality of the negro will retain the female slave.

I do not wish it to be inferred from this that I would treat as useless any efforts that may be made to put down this most shameful trade. I want merely to point out the difficulty, as I consider it, of its complete annihilation. The subject, however, is leading me away from matters of more immediate interest. So I resume my narrative.

I mentioned that the girls did not care to be set free, and were therefore allowed to follow their leaders.

I also prepared for my departure, compelled thereto more especially by the necessities of the stomach, which in journeys of exploration are just as imperious, and in fact even more so than geographical societies.

I therefore set out in an easterly direction, and after a two hours' march sighted a village and pitched my camp on the banks of a rivulet close up to it. I learned that both rivulet and village bore the name of Bembe.

When the work of cutting down the wood for our encampment commenced, I saw a sudden commotion among my blacks, who then took to their heels in every direction. Not understanding the cause of their panic, I immediately proceeded to the spot to make inquiries. On the very place which I had selected for my camp, appeared issuing from the earth millions of that terrible ant called by the Bihenos *quissonde*, and it was the sight of these formidable creatures which scattered my men. The quissonde ant is one of the most redoubtable *wild beasts* of the African continent. The natives say it will even attack and kill an elephant, by swarming into his trunk and ears. It is an enemy which, from its countless numbers, it is quite vain to attack, and the only safety is to be found in flight. The length of the quissonde is about the eighth of an inch; its colour is a light chestnut, which glistens in the sun.

The mandibles of this fierce hymenopter are of great strength and utterly disproportioned to the size of the trunk. It bites severely, and little streams of blood issue from the wounds it makes.

The chiefs of these terrible warriors lead their compact phalanxes to great distances and attack any animal they find upon the way.

On more than one occasion during my journey I had to flee from the presence of these dreadful insects. Occasionally upon my road I have seen hundreds of them, apparently crushed beneath the foot, get up and continue their march, at first somewhat slowly, but after a time with their customary speed, so great is their vitality.

It will not be out of place here to say a few words about some other African ants of a more ordinary kind than the quissonde.

One is a black ant, only half the length of the quissonde, but, like the latter, armed with powerful mandibles. The Bihenos call it the *olunginge*. It is the sworn enemy of the termites, against which it wages the fiercest wars, and generally comes off victorious, notwithstanding the smallness of its size.

These little ants are, however, a positive benefit to the natives, owing to the enormous havoc they make among the larvae, nymphae and eggs of the termites.

In certain places I found in the dwellings of the termites a large quantity of giant ants, some of which measured five-eighths of an inch in length, and which prey upon the abundant neuroptera of South Africa.

These ants, as I presumed, being but little given to build houses for themselves, take up their lodging with their more industrious neighbours.

None of these insects, with the exception of the quissonde, will attack man, but the latter will do so always, with the result of putting him to flight, as my carriers were ignominiously forced to do on the banks of the Bembe.

I had therefore to seek out another spot, as far removed as possible from the former one, on which to pitch my camp.

Some messengers whom I had despatched to the village of Bembe returned with the unpleasant news that the petty chief of the place had given orders to his people to sell me no provisions.

We were all beginning to feel the cravings of hunger; game there was none, and our entire food during the day had been a handful or so of massango, which fell to each of us in the division I made of what was obtained on the banks of the river Cuito.

The country in which we now stood was completely unknown to all of us, and as the natives without exception gave us a wide berth, we had no means of removing our ignorance.

I called my pombeiros together, and pointed out to them the absolute necessity of pushing ahead next day, in the hope of reaching a more hospitable region.

They agreed with me as to the necessity of doing so, and resolved to urge on their men as much as possible; no easy matter, however, on account of their being weakened through insufficiency of food. For the last two days I had observed vestiges of the country having been at one time exceedingly populous—ruins of old villages, some of them very old, being scattered here and there.

The questions arose, why had they been deserted? was the devastation due to slavery? was it owing to the insalubrity of the climate? was it caused by the dearth of game? Was it the inferior quality of the soil?

I could find no satisfactory solution to these queries; but the first suggestion appeared to me the most likely one.

Any way, this unexpected dearth of population caused us the greatest embarrassment; and as to myself, I that night suffered positive torture from the cravings of hunger.

Next morning, early, we had another mishap through one of the carriers falling ill, but my Doctor, Chacaiombe, though he could not cure the patient, nevertheless remedied the evil by shouldering the sick man's pack.

Just as we were leaving we had a visit from some of the natives—envoys of the chief of the Bembe—

who came to solicit something on his behalf. My only answer was a speech about what I thought of their dusky master, with which I sent them about their business.

I started at twenty minutes to nine. I had to wade across the river Bembe, which at that spot was 2½ yards wide and 3 feet deep, and was running in a S.W. direction into the Cuito.

The right bank was mountainous, but the left, after an almost vertical cutting 11 yards in depth, stretched out in a level and marshy plain nearly a mile in width.

The journey across, or rather through, the marshy ground cost us an hour, and was very fatiguing to the half-famished caravan.

The ground afterwards appeared slightly inclined and covered with scrub, which was very difficult of passage. After a trying march of another hour or so, we came to an incline, at foot of which appeared a plain whose extent was incalculable, owing to a dense forest. We descended for some 60 yards or so till we reached the edge of the wood, but had then to alter our course, as the jungle was simply impenetrable.

We lighted upon the track of some animal, which we followed now eastward, now north-westward, and then south, until we came to a dead stop on the edge of a precipice, 300 feet at least in depth, at foot of which was brawling a mountain torrent.

The difficulty of the path, the heavy loads with which the men were weighted, and the weakness of the latter, induced me to call a halt and pitch our camp.

The hunger from which we were suffering was beginning to get unbearable. There was but one hope which animated me, namely the having seen vestiges of game.

It was not a pleasant circumstance that immediately upon our arrival we should be visited by a cobra, which we fortunately killed. My Doctor averred that it was of a most poisonous kind, but that he possessed an antidote to its bite. In spite of this assurance I felt more confidence at seeing it most effectually put out of the way. It was upwards of a yard in length, of a dusky red upon the back and of a lighter tint upon the belly. The eyes were green and brilliant as emeralds, and it had a bipartite tongue.

The mouth was armed with four teeth disposed like those of a dog. I mention these features as they may be serviceable to those who follow me on this road.

I felt that game must be obtained, for nature could not hold out much longer. Having therefore made my arrangements, I started off in one direction and sent my attendants Augusto and Miguel, the only trustworthy woodsmen I possessed, in another.

Shortly after leaving the camp I found the track of a herd of buffaloes, and at once followed it.

I may here mention incidentally that few sportsmen in Europe can form an idea of what it is to hunt for actual food. If it can be called a pleasure it has a good alloy of pain in it.

It may be likened in some measure to the mixed feelings of a gambler who approaches the table for the purpose of gaining the wherewithal to pay a debt of honour, and who, while delivering himself up to the feverish joy of play, has in his heart the tearing anxiety of uncertainty. The eyes of the man, while devouring the cards which slowly fall from the banker's fingers, and seeking to penetrate through the bits of cardboard, the quicker to remove his agony of doubt as to whether it is safety or destruction they will bring to him, must surely wear some such expression as that of the half-famished huntsman who follows the trail of an animal whose possession is to him a matter of life or death.

There, however, the resemblance between them

ceases; for the huntsman can at least, in singleness of heart, invoke Divine assistance in his quest and his success.

How different again from the sensations of the hunter stimulated by hard necessity, are those of the sportsman whom pleasure only brings into the field! However keen may be the zest with which the latter pursues his prey, he is not so indifferent to surrounding objects but that he can stop awhile to admire the lovely landscape or pluck a brilliant flower, conscious as he is that if he fail in bringing down his quarry, his table will be none the less well supplied, or his bodily comforts be less carefully ministered to.

The other man sees nothing, hears nothing, but that has a bearing upon the one desired object. Heedless of the thorns that tear him, or the boughs that bruise him in pushing them too carelessly aside, with set teeth and beating heart, his empty stomach egging on his waning courage, he pushes forward to reduce to the utmost the distance lying between him and his prey, the surer and the deadlier to make his aim.

The trail I followed led me at length to the very bottom of the precipice, where the water was brawling over its uneven bed, and for a considerable time I kept along its right bank, till, finding an opportunity, I crossed to the other side, whence I perceived my buffaloes grazing at the outskirt of a dense virgin forest.

They were at least 500 yards from me.

Then began the fatiguing operation of stalking—my gun on the trail, wading, as it were, through a sea of dry grass. From time to time I would raise my head to see how much my distance had been shortened, and to make sure the creatures had not taken the alarm. The very idea brought the moisture to my skin, for I longed in fancy to return to the camp and bid my followers hie to the banks of the torrent, where they would find provision to stay the cravings of their hunger.

My hopes and fancies were dispersed as if by an enchanter's wand. When I lifted up my head for the last time, not a buffalo was visible. They must have disappeared within the forest.

I rose in all haste and, with the utmost speed of which I was capable, followed in the direction they must have taken. It was perfectly in vain. The thick and springy moss which covered the ground left not a trace of their passage, nor could my keenest endeavours overcome the difficulty.

It was a deep disappointment, unrelieved by any after-success; so that about six in the evening, worn out with fatigue and hunger, I made my way back to the camp, having, as I calculated, covered some twelve miles in vain.

The others, however, had been more happy than myself. Augusto came running out to meet me with a radiant face, and with no little triumph led me up to a superb antelope which he had shot a little while before. It was an enormous *Hippotragus equinus*, as bulky as an ox.

I lost no time in cutting it up and dividing it equally among us all; and after so lengthened a fast, which, as being most involuntary, I am afraid cannot be placed to the credit side of my account hereafter, I made such a meal as only those who have been in the same fearful straits can adequately appreciate.

The contentment caused in my whole being by the consumption of a hearty supper was somewhat dashed at the aspect of my worthy Miguel, the elephant slayer, who appeared before me with such a long face that I was sure something very serious had occurred to disturb him, and when I learned the cause, I did not wonder, though I could not help being inwardly amused, at his dismay.

During his absence my pet goat Cora had got into his tent and sacrilegiously munched up the wonderful charm which he possessed for slaying elephants!

This marvellous talisman consisted of a human tooth fallen from the jaw of some antiquated skull, wrapped up in straw and rags by a medicine-man of high repute who had imbued it with sovereign virtues, so that the possessor of the treasure would find it easy to fall in with and slay elephants without the slightest danger to himself. Miguel was for a time inconsolable, but I managed at length to pacify him by the promise of a far more effective charm than the one he had lost.

Nor did I deceive him in making such a promise, since the excellent rifle I intended to bestow upon him, when we reached the elephant country, would, I conceived, be of far greater value than any amount of rotten teeth packed up in rags and straw.

After our meal, my pombeiros gathered about my fire, and, amid other things, related that, during my absence, the men had wandered about the wood, where some had collected a lot of honey, and others had gathered quantities of a fruit which the Bihenos called *atundo*, and which grew upon a stunted herbaceous plant. The fruit-stalks spring from the stem quite close to the ground, and the fruit is just as much below as above the earth. It is agreeable enough to the palate, but I doubt it being very nutritive.

It was necessary to be on the move betimes the following morning, so that we broke up our camp much earlier than usual, in spite of the cold.

We started in a S.E. direction, and after two hours' march, came upon a river that was very difficult to cross over. Its width was rather more than four yards, and depth quite as much, with a violent current.

I gave orders for the felling of some large trees and

managed to throw them across the stream by way of bridge, over which my entire caravan passed in safety. A little below the spot where we crossed the stream, it receives the waters of a rivulet running from the eastward I followed the course of this rivulet on its right bank for upwards of an hour, and subsequently halted near two villages.

Immediately upon our arrival several of their inhabitants gathered about us, with whom we had a parley about provisions. A lot of massango—the canary-seed

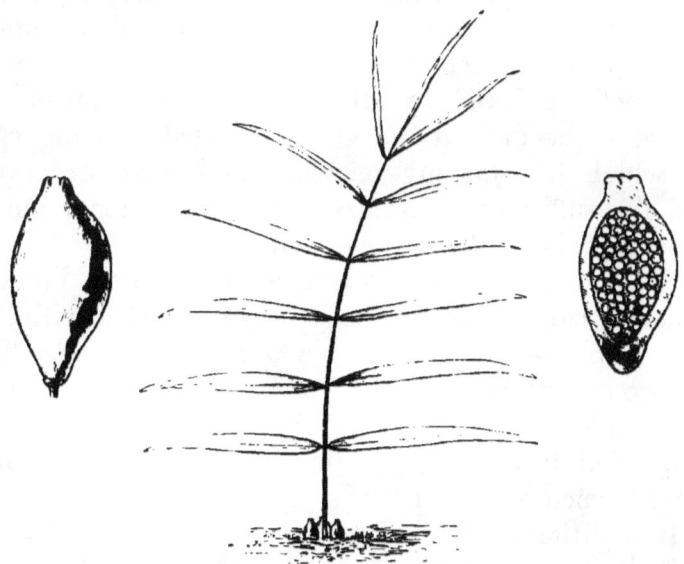

Fig. 47A.—ATUNDO, PLANT AND FRUIT.

before alluded to—was soon brought into camp by negroes who were almost entirely destitute of clothing, and, as we did not dispute about price, we shortly had sufficient for that day's consumption.

Friendly relations were soon established between the natives and my people. I learnt from them that the rivulet by which we encamped the evening before was called Licócótoa, the one over which we had thrown the bridge, Nhongoaviranda, and the brook by whose sources we were now staying, Cambinbia.

The two villages built upon the left bank of the little stream were Luchazes, but that to the N.W. of my camp was inhabited by Quiôcos or Quibôcos. They were the latter with whom we were in communication.

I consumed more than a pint of massango boiled in water, and did not find it an unpleasant food.

After satisfying my appetite, I calculated the position in which the planet Jupiter would be that night at the time of the eclipse of the first satellite, which I wanted to observe, but my camp was pitched in a dense forest which prevented me seeing the stars.

Directly I found by calculation the position of the planet at the desired moment, I selected a fitting spot on which to plant my telescope and gave orders to make a sufficient clearance of the wood about me to allow my observations to be made.

The tangled jungle was tremendously thick, but my Bihenos, hatchets in hand, set to work with a will, and in a couple of hours gave me a clear opening. The Quiôco or Quibôco women who visited our camp carried their children by their sides like the Luchazes, that is, suspended from the opposite shoulder by means of a sling formed of the bark of a tree.

In addition to the massango, they brought with them for sale certain tuberculous roots called *genamba*, which my people seemed to enjoy immensely, but which I found anything but agreeable. They do not grow maize, and feed almost entirely on massango.

The extravagant head-dresses to which I have more than once alluded are not observable among the Quibôcos or Quiôcos, and their body covering is more miserable than that adopted by the Quimbandes. The women, as usual, are more scantily clad than the men.

My readers may, perhaps, feel some surprise to hear

me talk of Quiôcos when I am in the very heart of the Luchaze district, and I can assure them that my astonishment was quite as great at finding them there.

The constant emigration of the Quiôcos and the colonisation by them of the Luchaze territory are undoubted facts.

The country of these Quiôcos or Quibôcos (for they are called indifferently by both names) is situated to the north of Lobar, on the eastern slopes of the Serra da Mozamba. Livingstone makes it cut by parallel 11 south, and by the 20th meridian east of Greenwich.

The Quiôcos are travellers and bold huntsmen. Many of them, dissatisfied with their own country, emigrated southwards, crossed the Lobar, and established themselves on the right bank of the Lungo-é-ungo in the Luchaze territory.

Finding themselves unmolested, they were soon followed by others, so that at the present day the emigration is constant. They have not, however, all stopped there, many having gone still farther southwards and settled on the banks of the Cubango. The greater part of the inhabitants of Darico are Quiôcos.

In answer to my inquiries as to the motives which induced them to leave their country, they said it was sickness and the scarcity of game.

The Quiôcos with whom we were in communication were only recent settlers and had no store of provisions to dispose of; but they informed me that on the other side of a lofty serra lying to the eastward, there were several Luchaze villages and abundance of food.

I hired guides to take us thither and resolved on starting the very next day, which, however, I was prevented doing by the illness, during the night, of several of the men.

My young nigger Pepeca appeared before me in the morning with an enormous goître-like swelling, and almost all my people were suffering more or less from the stomach, no doubt owing to the massango they had eaten, but to which they got quite habituated later on. Happily, I myself felt no inconvenience from the new kind of food.

I sent to the two Luchaze villages on the left bank of the Cambinbia, but my messenger returned empty-handed, as the natives refused to sell them anything. We owed to the Quibôcos the provisions we required for that day's consumption.

There were some of the men on the sick list the next day, but we were compelled to leave, as the natives pointed out the impossibility of furnishing me with anything more to eat. I obtained from them a few men to supply the place of the carriers who were invalided, and at nine next morning we left the camp and, preceded by the guides, directed our course to the Serra Cassara Caiéra, the lofty mountain to which allusion had been made the day before, and beyond which we were to find abundance of provisions.

The actual height of the mountain is 5298 feet above the sea-level, or 450 feet above my camp on the Cambinbia. It forms a table-land with tolerably steep slopes. The climb to the top was fatiguing. During the process the carriers beguiled the time, and perhaps lightened their labours, by a monotonous chant, which literally translated ran as follows:

"The cobra has no arms, no legs, no hands, no feet. And yet he climbs the mount! Why should not we get up as well, with arms and legs and hands and feet?"

I went on for about an hour along the summit of the serra from west to east until I came to the descent.

From the highest point, a magnificent panorama meets the eye of the spectator, extending from N.E. to N.W. The entire course is visible of the river Cuango,

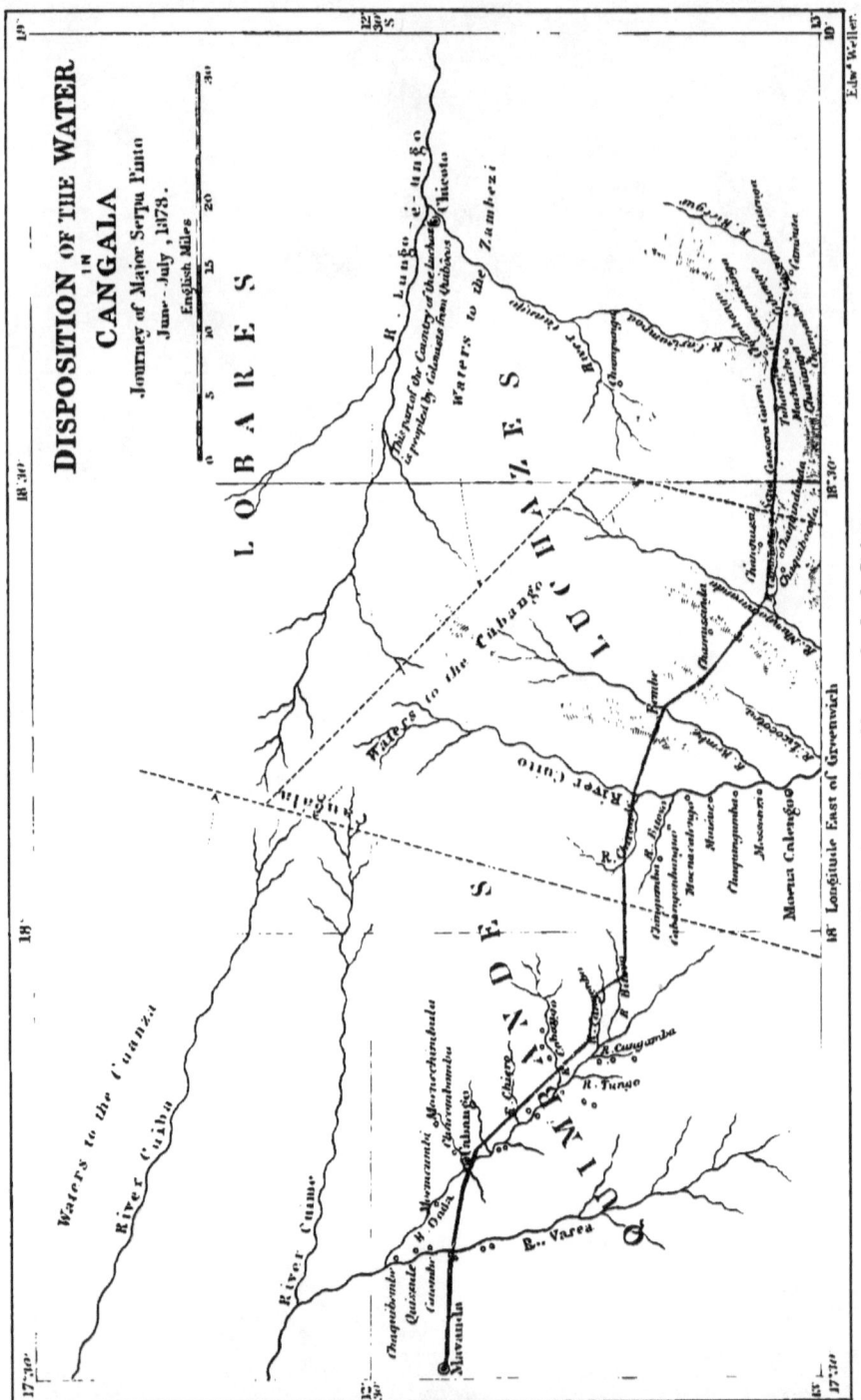

the southern affluent of the Lungo-é-ungo. The eye can distinguish the vast watershed of the latter river from Cangala to the confluence of the Cuango, together with the higher watersheds of the rivers Cuito, Cuime and Cuiba. The extent of prospect is truly surprising.

On the western slope of the *Serra* the arboreous vegetation is splendid; the summit, which, as I have hinted, is of considerable length, is somewhat poor, but the eastern slope again is wonderfully rich in trees and shrubs. This eastern side bears the name of Bongo-Jacougouzêlo.

I pitched my camp at the source of the Cansampoa, a rivulet which runs into the Cuango; not having met with a drop of water throughout the day's journey.

In the immediate vicinity of my camp, but on the other side of the rivulet, were five Luchaze hamlets. They are all governed by a petty chief who does homage to the Sova Chicôto, whose village is situated at the confluence of the Cuango with the Lungo-é-ungo. The other two Luchaze hamlets, which are on the Cambinbia, are subject to the Moene Calengo on the river Cuito.

The petty chief before alluded to, and who rejoices in the name of Cassangassanga, came to call on me, bringing with him a kid by way of present. A few beads made him quite satisfied. He promised to send me some massango, and guides to conduct me to the village of Cambuta, where, as he alleged, I should find abundance of provisions. He was as good as his word, and the massango and guides appeared in due course.

The massango, when divided, supplied a scanty ration to each of us, and, as the kid was but a small one, we went to bed with appetites very inefficiently satisfied.

The natives cultivate massango, a little manioc, still fewer beans, the castor-oil plant, in tolerable quantities, and a few hops.

They work in iron with considerable skill, the ore being found in the country.

On the 6th of July I started in an easterly direction, and, after three hours' journey, the last of which was along the bank of the Causampoa rivulet, I camped near the river Bicéque, which runs in a

Fig. 48.—Village of Cambuta, Luchaze.

N.E. direction to unite its waters with those of the Cutangjo, an affluent of the Luugo-é-ungo. The country is dotted over with hamlets, whose populations obey the Sova of Cambuta. I was able at the latter place to get a tolerable supply of massango, the sole article of food they cultivate in any quantity, and consequently the only one they offered for sale.

Fortunately there were large flocks of wood-pigeons: indeed, I never saw them more abundant than in this district; and I managed to bring down not a few, charging my gun with little pebbles from the bed of the rivulet.

At this time, several of my carriers fell ill; some suffering from goître, and others from inflammation of the stomach, arising doubtless from bad and insufficient food.

Among the girls who came into my camp to dispose of massango I noticed more than one of elegant form and graceful carriage. It could not be said that they

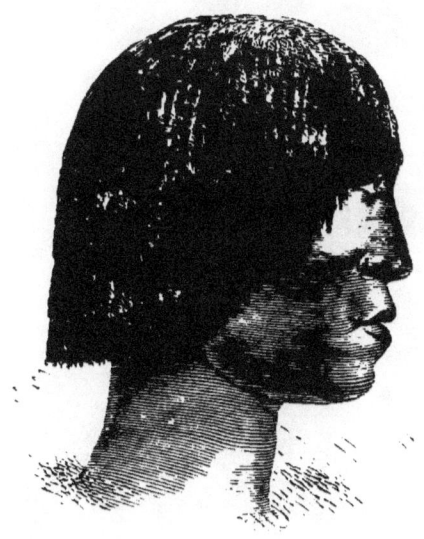

Fig. 19.—LUCHAZE WOMAN OF CAMBUTA.

owed anything to art, for clothing they had none; a little strip of the bark of a tree doing duty for the traditional fig-leaf.

I further observed that both men and women, without exception, had their four front incisors fashioned like a triangle, so that, the teeth being closed, there appeared a lozenge-shaped aperture in the middle.

The cold continued intense during the night, and

we could get no rest except in the neighbourhood of our fires.

On the following day there appeared greater sickness than ever in camp. It is worthy of remark that the Bihenos only were attacked, whilst the Benguella negroes, who are far less inured to the exposure and vicissitudes of travel, almost entirely escaped.

In the morning a large and ferocious bird was killed

Fig. 50.—LUCHAZE MAN OF CAMBUTA.

in the neighbourhood of the camp. It might have been want of special knowledge on my part, but I could not assign it to any of the kinds into which the family of diurnal birds of prey are divided. I considered it a species of vulture, although solitary of its kind. It certainly greatly resembled the vulture, saving that its dimensions were somewhat smaller, measuring but 3 feet 9 inches from wing to wing.

Vulture or not, it was a *bonne-bouche* to my Bihenos,

to whom, in matters of gastronomy, nothing came amiss, from their fellow men to cormorants, with crocodiles, leopards and hyenas in between : all were welcome to their insatiate jaws.

On that day, as the day before, every hour that I could spare from my observations I spent in scouring the neighbourhood and drawing up a rough map of the

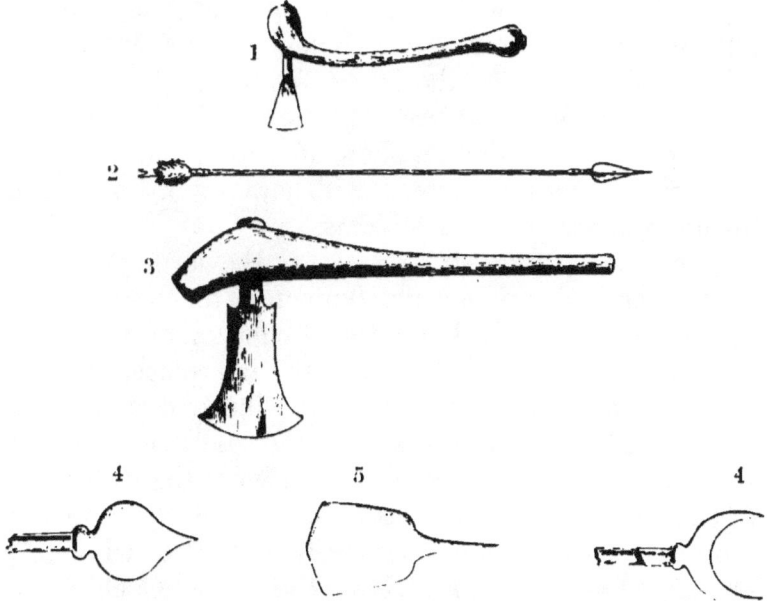

Fig. 51.—Articles manufactured by the Luchazes.
1 and 3 Hatchets.   2. Arrow.   4, 4. Arrow-points.   5. Spade.

district, taking in three miles to the south of the source of the Biceque, which area comprised the source of another river, the Cuanavare, a great affluent of the Cuito. Near the source of the Cuanavare I came upon the village of Muenevinde, governed by a woman, whose husband, by name Ungira, had no active voice or part in the government.

I never could be said to be passionately fond of

kidney-beans, but that night, on my return to camp, I had a small present of these dainties made me, and, truth to say, I devoured them with infinite appetite.

The Sova of Cambuta was absent at the chase, and the honours of his house were done me by his wives, with whom I was soon on the most cordial terms. I obtained from them not only a good share of massango, but a dozen porters to carry it, and two guides to lead me to the sources of the Cuando and the Cubangui, an affluent of the latter—rivers which the natives of the country told me were the *largest in the world*.

Grandiloquent as this designation undoubtedly is, it is not bestowed without a show of reason, and, with the permission of my readers, I will here say a few words about these magnificent streams.

The river Cuando, of a certainty the largest affluent of the Zambesi, was not first known to me through the information furnished by the Luchazes of Cambuta. In my journey from the Bihé to that place, I kept much more to the north than the Biheno caravans are accustomed to do; and this I did purposely, fully aware that sooner or later I should fall in with the watershed of that great artery. I was influenced in the course I took by conversations with Silva Porto, who had already descended that river from the Cuchibi to Liniante, conveying goods in canoes.

He had furnished me with certain data as to the sources of the river (with the central and lower portions of which he was personally acquainted), and which, from information supplied him by the natives, he fixed at very nearly the spots where I actually found them.

If Silva Porto could only give to the places that he knows in South Central Africa their correct positions in latitude and longitude, many of the blanks that now exist in the maps of the country would speedily be filled up.

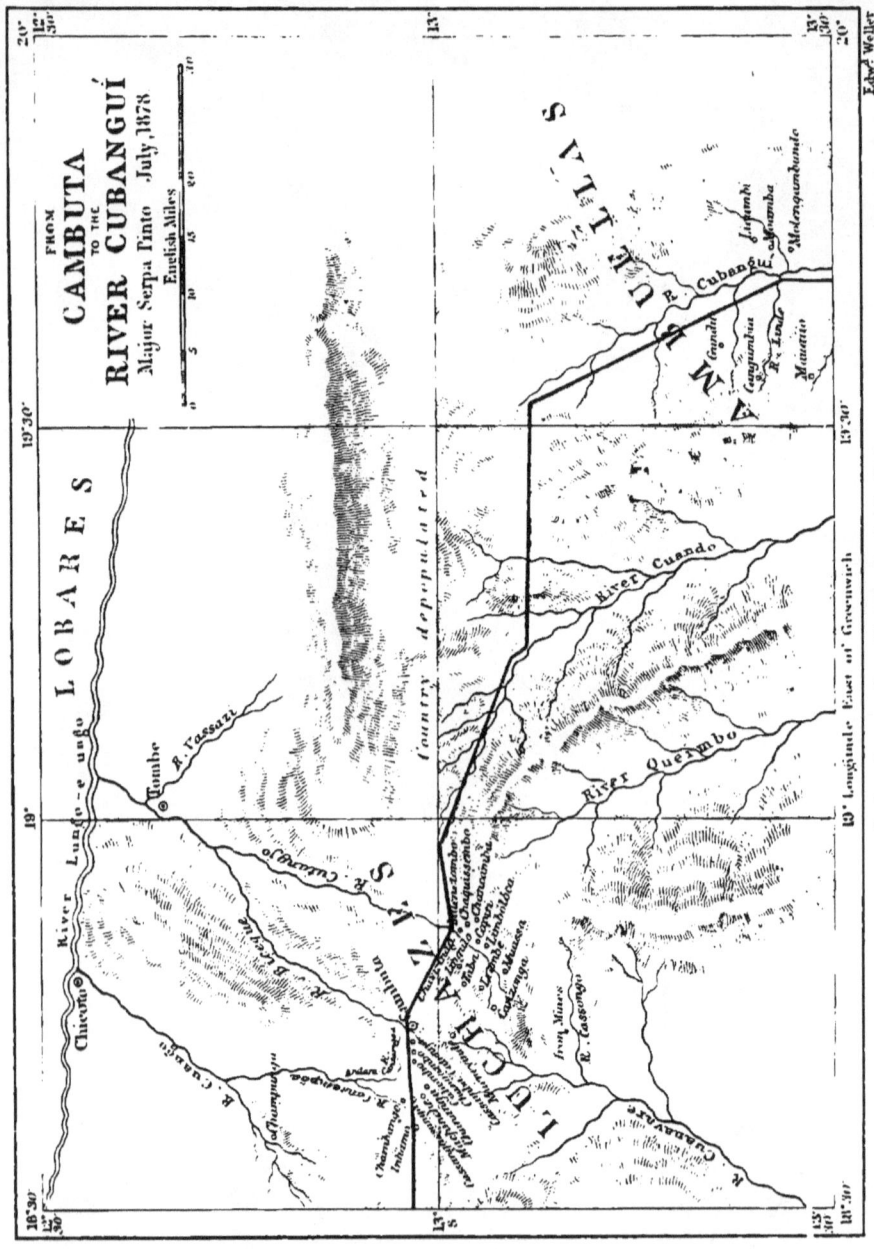

On leaving Cambuta, therefore, in search of the sources of the Cuando, I was only completing the itinerary I had traced out, and was endeavouring to solve one of the problems that I most ardently desired to unravel. As I went on, I was collecting at every step interesting matters of detail; the general features having already been delineated by Silva Porto.

My guides had informed me that we should have to traverse, beyond the river Cutangjo, a waste and unpopulated region, so that it behoved us to provide ourselves with ample stores for the journey. It was this communication which led me to purchase so large a quantity of massango, and to hire of the Sova's wives twelve men to carry it.

I started on the 9th of July, at 9 o'clock in the morning: three hours later I crossed the river Cutangjo, and camped on the right bank of that river near the village of Chaquissembo. The Cutangjo is there 4½ yards wide by 3 feet deep, and runs N.N.E. towards the Lungo-é-ungo. I observed that in the plantations there was some manioc and a great deal of massango—that terrible massango, which literally haunted me in Africa!

The Luchazes cultivate, to some extent, the cotton and castor-oil plants. They work the iron which they obtain from the banks of the Cassongo, and are very skilful smiths.

Almost all the Luchazes are furnished with a beard beneath the chin and a small moustache. But the extraordinary fancy in head-dress, which I have more than once referred to as exciting my wonder and admiration, is unknown among them.

The men wear a broad belt of untanned leather, fastened with buckles of their own manufacture; they cover their nakedness with skins, and further shelter themselves from the cold with *licondes*, a rough kind of cloth woven from the bark of various forest-trees.

They make no pots or pipkins, and those they use are obtained by barter from the Quimbandes.

They fashion bracelets out of copper, which is supplied by the Lobares in exchange for wax, the Lobares themselves obtaining the metal from the Lunda.

I paid a visit to the village of Chaquicengo, which,

Fig. 52.—Luchaze Woman of Cutangjo.

like the whole of the inhabited places throughout the country, is very pretty and extremely neat. The houses are made of the trunks of trees, about 4 feet in height, which is in fact the height of the walls. The space between each upright is filled in, occasionally with clay, and in other cases with straw. The roofs are thatched, and, as the frame-work is composed of very fine rods, the thatch bends inwards and produces an effect similar to the roofs of the Chinese. The granaries are perched at a considerable height upon a timber frame-work, entirely of straw, and a movable cover, so that it is necessary to remove it to get inside to seek for stores.

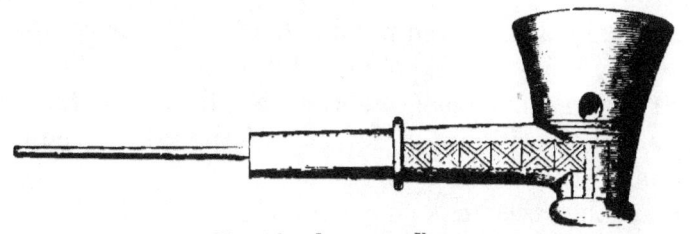

Fig. 53.—LUCHAZE PIPE.

Access is obtained by means of a hand-ladder, and they are, in fact, little more than gigantic water-proof baskets, on which conical covers have been placed.

Fig. 54.—LUCHAZE FOWL-HOUSE.

The fowl-houses are quadrangular pyramids of twigs of trees, placed upon four lofty feet or stakes, to protect the inmates from the attacks of small carnivora.

In the centre of the village I observed, as in the

Cuambo, a *Kiosque* or temple for meeting or conversation.

I found several men squatted round the hearth, busy making bows and arrows. They received me very courteously, and offered me for drink a liquor composed of water, honey and powdered hops, which they mix together in a calabash, where it is allowed to ferment. They called it "bingundo," and I thought it the most alcoholic stuff I had ever tasted.

The Luchazes make use of a gin or trap to catch small antelopes and hares. It is ingenious in con-

Fig. 55.—The *Urivi*, or Trap for small Game.

struction, and will be readily understood by a glance at the drawing. The name they give it is *Urivi*.

On my return to camp, after an excursion to the sources of the Cutangjo, I was accompanied by a large number of men and women, who were never tired of looking at and watching me. They were none of them remarkable for beauty; but, on the other hand, among the natives on the banks of the Cutangjo, I saw not a few male specimens of perfectly revolting ugliness.

These people not only collect a great deal of wax from the forest, but encourage the bees by furnishing them with hives, formed of bark and strips of wood, which they fasten in the branches of the trees.

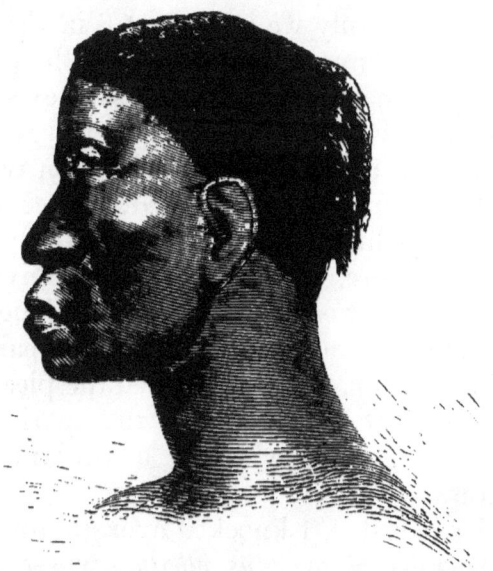

Fig. 56.—Luchaze of the Cutangjo.

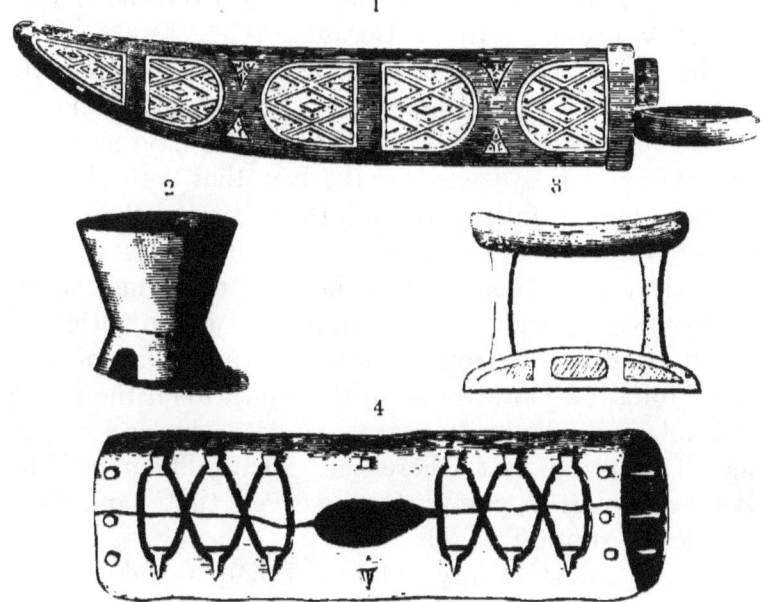

Fig. 57.—Luchaze Articles.
1. Knife-sheath. 2. Basket. 3. Wooden bolster. 4. Bee-hive.

On the 10th of July we started at 8 in the morning, and half an hour later, notwithstanding the presence of the guides, we lost ourselves in a forest of excessive density, from which we only managed to emerge, with considerable trouble, at 10 o'clock. We then traversed a space that was free of underwood, but covered with gigantic trees, which shaded us delightfully from the sun. This pleasure, however, was short-lived, for in another half-hour we were in a thick jungle again, where locomotion was difficult and even painful. At last, at 20 minutes past 11, I descried the pleasant slope of an eminence, at whose feet lay the sparkling water of a little lake, surrounded by a verdant carpet of waving grass.

Just as I reached it, I knocked over an animal, which I believe is called *Leopardus jubatus*, whose skin went to swell the number which constituted my feline bed. This skin, on which I slept as far as Pretoria, I subsequently presented to Dr. Bocage.

The *Leopardus jubatus* must be rare, as I only saw two specimens throughout the course of my journey. Its sight, I presume, is defective in the daytime; my supposition being based on the fact that both of those animals, on my falling in with them, turned their ears, rather than their eyes, in my direction, as if they trusted more to their sense of hearing than their sight.

Having determined the position of the sheet of water, I drew off from it, and had my camp pitched some hundred yards or so to the south upon the rising ground, and about 90 feet above the surface of the marsh; for the spot where the great affluent of the Zambesi takes its rise rather deserved that name than the designation of a lake.

In the midst of my labours I had a sudden and violent attack of fever, which completely prostrated me for some three hours. When I came to my senses, I

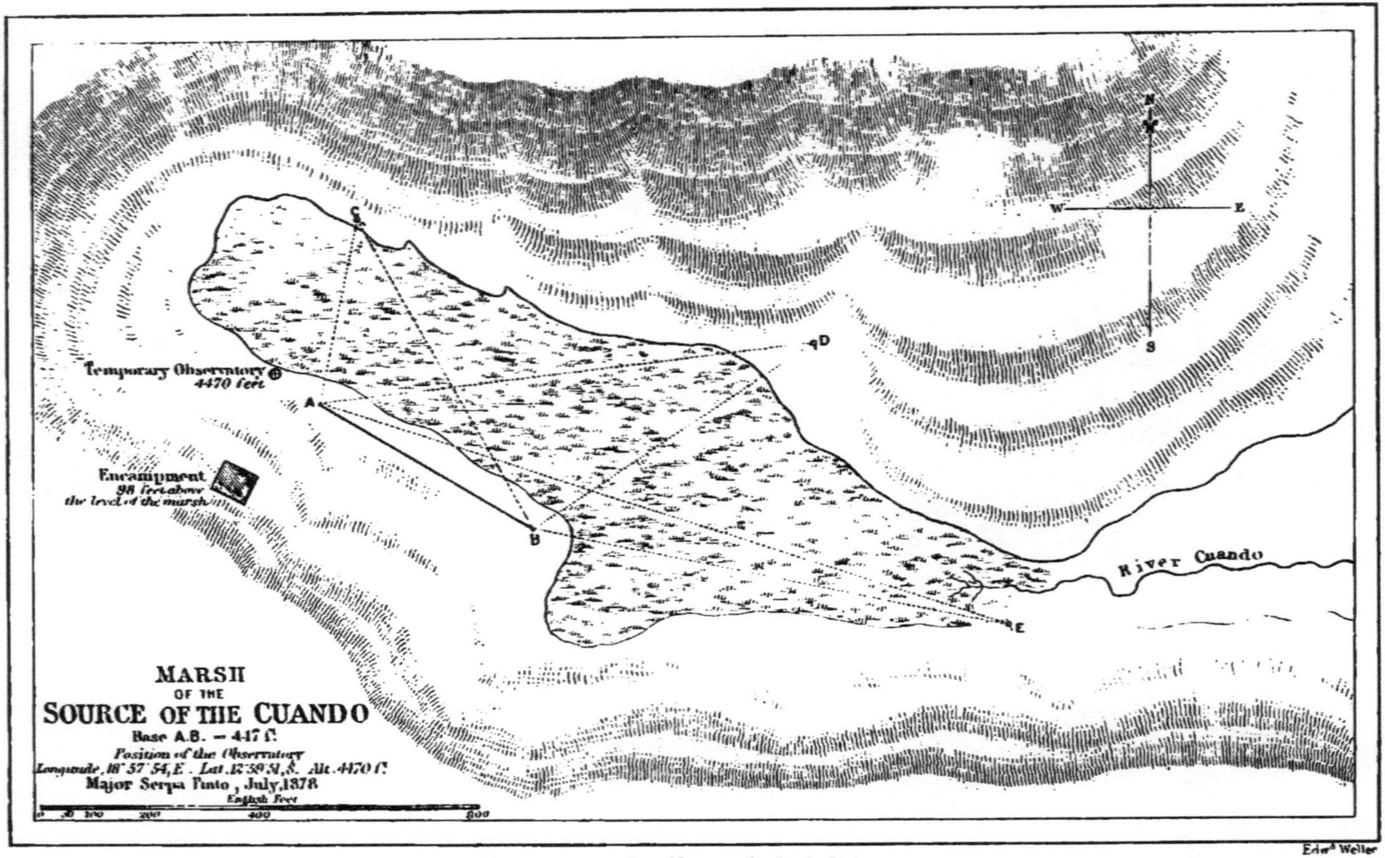

London: Sampson Low, Marston, Searle, & Rivington.

could scarcely refrain from smiling at my curious plight. I was literally covered with amulets, my chest alone being thickly strewed with the horns of small antelopes, full of the most precious medicines. A bracelet of crocodiles' teeth encircled my right arm, and two enormous buffalo-horns were suspended from a couple of poles set upright in my tent.

During the fever, my negroes had lavished the greatest care upon my person, and, in obedience to Dr. Chacaiombe's instructions, had heaped these things upon me with the utmost faith in the result.

A strong dose of quinine, which I took as soon as I was able, brought about my speedy recovery, a result that was no doubt, however, set down to the virtues of the amulets.

My attendants, Augusto and Miguel, went out upon the hunt for game, but brought nothing back with them, though they sighted a few leopards in their rambles. Many traces of larger game were, however, visible on all sides.

Early next day I drew up a rough map of the marsh; rectified my position, and constructed a small monument of clay in the hut where I made my observations. Within this tumulus I buried a bottle which had contained quinine, carefully wrapped up, and containing a paper, on one side of which I wrote the names of the members of the Central Geographical Commission, headed by that of His Majesty the King of Portugal, and on the other the co-ordinates of the spot and the date.

After mid-day the Luchaze guides took me to see the source of the river Queimbo, an affluent of the Cuando, on the western side. I set it down at 6 geographic miles S.W. of the marsh, forming the source of the Cuando itself.

My twelve Luchaze carriers were very home-sick,

and complained bitterly of the cold. The country is depopulated, and should contain a great deal of game, judging from the traces that were observable. Another clear evidence of the fact was the number of leopards we started, but, unfortunately for us, we started nothing else. And we could not afford to linger, for our provisions were rapidly disappearing, and our only chance of relieving our hunger was to reach the Ambuella villages without delay.

On the morning of the 12th July, with a temperature only 2 degrees above zero, I broke up my camp and prepared to leave, though we did not make a start before 8 o'clock.

Thousands of paroquets, that were harboured in the woods, were all shrieking at once, and the noise they made was perfectly deafening.

I kept along the right bank of the Cuando for a couple of hours, and then, at the direction of the guides, crossed over to the left, by a bridge which we improvised out of the trunks of trees.

The river measured there between 6 and 7 feet wide, and about the same in depth, with an excessively rapid current.

I was just crossing the river, when I observed a herd of gnus, at which, however, I could not get a shot.

I encamped beside the river. The banks of the Cuando are mountainous, and from its source to that point they are flanked by marshy ground, some 30 to 40 yards wide, yielding abundance of water, which drains into the river.

This peculiarity is noticeable with almost all the rivers of those regions, which thus receive enormous quantities of water, so that, even without any subsidiary streams, they become navigable at only a few miles' distance from their unpretending sources.

I observed on the right bank of the river, both at that spot and others, several vertical stratifications of a red, white, and azure colour.

The next morning, at 8, I was again on the move, and tramped on till noon, camping at that hour near a brook which ran into the Cuando.

Fig. 58.—The Cuchibi.

I had some of the men on the sick-list, a few suffering with goître, and others with inflammation of the legs.

Happily, for their sake, the loads of provisions had sensibly diminished, and I had now carriers to spare. The marshy banks of the Cuando abounded in leeches, and I had a lot of them caught to apply to such of my patients as stood in need of their assistance.

The woods I had passed through, and the one where I was now encamped, were almost exclusively composed of enormous trees, which the Bihenos styled *Cuchibi*, and that turned out most serviceable to my half-famished caravan. They produced a fruit not unlike a French bean, having one bright scarlet seed enclosed in the dark-green husk. After a lengthened decoction, the scarlet envelope separates from the white sheaths and forms the edible portion of the fruit. These seeds are very oleaginous, and both the Ambuellas and

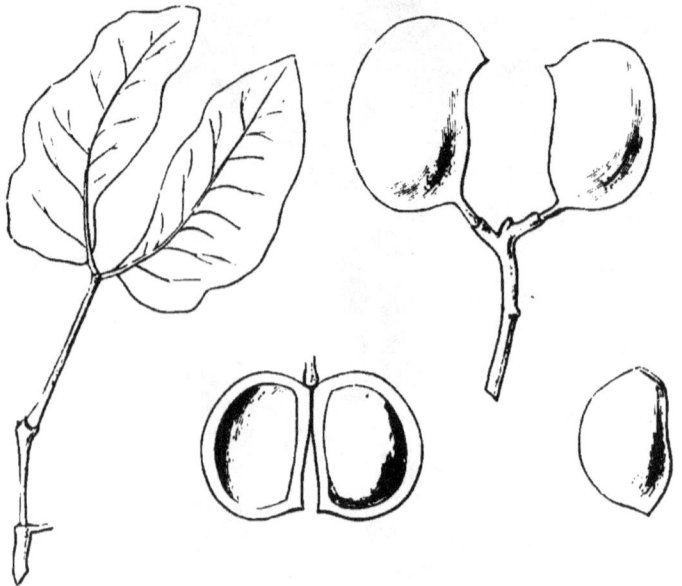

Fig. 59.—Leaf and Fruit of the Cuchibi.

(Natural size.)

Luchazes extract from them the oil with which they moisten their food.

This fruit is undoubtedly a great resource to the hungry traveller, but it is of no use to the hurried one, as the decoction is a work of time.

There is also another fruit in these parts, and which is exceedingly common. The Bihenos call it *Mapole*. It is the product of a tree of medium size, the *Mapoleque*,

AMONG THE GANGUELLAS.

and resembles an orange both in colour and dimensions. It hangs vertically from the branches of the tree, suspended from a longish stalk. The outer rind and its lining, closely adhering to each other, form a husk about an eighth of an inch in thickness, and hard as horn. It can only be broken with a strong hatchet.

Fig. 60.—The Mapole, Tree and Leaf.

When opened, it displays a thick and coagulated liquid, full of seeds similar in size and appearance to the stones of small plums.

This liquid, of an acid-sweet taste, taken in any quantity, is very purgative; but the Bihenos assured me that it was most nutritive, and would support a man for a day or two easily enough. I did not myself put it to a personal test.

The next day I left the river Cuando, which already at that spot inclines to the S.S.E., and, in obedience to the orders of the guides, travelled eastward in search of the sources of the Cubangui, a river which they asserted to be very large.

After an hour's march, I passed a brook running to the south, through marshy land, a hundred yards or so in width, which it cost some labour to traverse. Four miles further on I came upon another large brook, running parallel to the previous one.

Between the beds of these brooks, as well as between those of the affluents of the Cuando to the eastward, is a chain of mountains lying north and south. These mountains belong to a more important system, which to the north lies east and west, and terminates in the valley of the Lungo-é-ungo.

At about half-past 11 I arrived at the summit of the Serra, whence the guides pointed out to me, in the far distance, the sources of the river Cubangui. I could define them perfectly to the eastward, but as I could not, immediately on my arrival, determine the latitude, I took a rest, and at noon fixed the latitude of the point where I stood, which was the same as that of the sources of the river; the two lying due east and west of each other.

At 2 in the afternoon I camped hard by the sources themselves, and found them to be similar in character to those of the Cuando. The axis of the marsh in which the river takes its rise lies north and south, and extends three-quarters of a mile in length; the width varying from 80 to 100 yards.

No game was visible, but many traces of it appeared, and during the night the lions all about the camp kept up a horrible concert.

Our last rations were here served out, and hunger again stared us in the face.

The guides averred we were at no great distance from the villages, but it would take us at least a couple of days to reach them, owing to our numerous invalids, and more especially a pombeiro of the name of Canhengo, who was very ill; forced marches, under the circumstances, were therefore out of the question.

I felt excessively anxious, my great fear being that

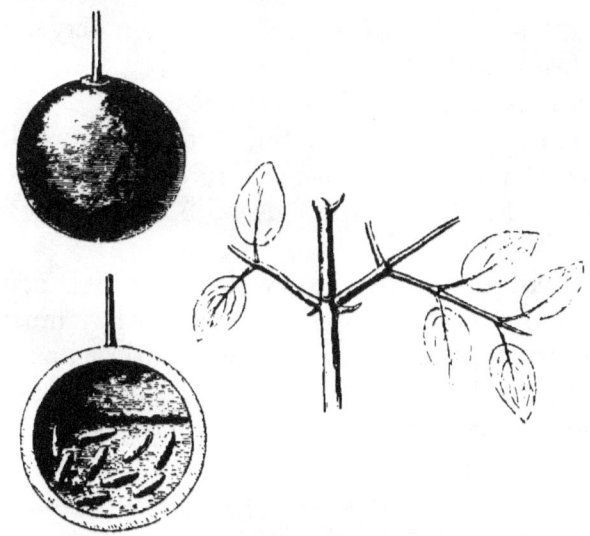

Fig. 61.—MAPOLE, FRUIT AND ARRANGEMENT OF THE BRANCHES.

hunger and fatigue would so aggravate the condition of the sick as to prevent me obtaining the requisite resources for us all, in time to be of any use.

On the following day, in spite of all my efforts, I could not keep up the march of the caravan over four hours, and was compelled to camp alongside the Cubangui, which river, in fact, I had not left from the time of making its source. At the place where I came to a halt it was already 3 yards wide, and between 3 and 4 feet deep.

A gnu which I shot, and a little honey which the negroes gathered in the forest, furnished our only rations for that day.

Next morning I went on again, following the right bank of the stream, and after another 4 hours' march camped beside the Linde rivulet, opposite three Ambuella villages. I at once despatched messengers, not only to those places, but to others lying on the same side of the river as ourselves, but all we obtained was a scanty supply of massango. We, however, received the information that another day's march would bring us to the lands of the Sova, and that from the latter we could obtain provisions.

I found by measurement that the Cubangui's dimensions were increasing apace. Here, at the confluence of the Linde, it was upwards of 5 yards across, and was 9 feet deep.

My invalids improved very slowly, which could not be attributed to want of strict diet. They must have suffered too not a little from actual fatigue, as it took 6 hours next day to reach the Sova's village, Cangamba. I forthwith despatched a present to the great man, in the shape of an old uniform of an infantry captain, with which he was delighted, and gave prompt orders to his people to supply me with food. We obtained, in exchange for beads, some of that eternal—I had almost said cursed—massango, from which there appeared now to be no escape.

I discharged my guides and the 12 Luchazes who had accompanied me thus far, and who took their leave well satisfied with what I gave them.

They fraternised easily with the inhabitants of the Ambuella villages, which are in fact partly peopled by Luchaze natives. Of this I had a clear proof a day or so after my arrival, when several Luchaze families, who had emigrated from their own country to establish themselves in this district, pitched their camp within a stone's throw of mine.

I saw and conversed also with a band of hunters who were travelling southwards in search of elephants. It was the first time I had heard speak of elephants, as not one is to be found throughout the country I traversed

Fig. 62.—MOENE-CAHENGA, SOVA OF CANGAMBA.
1. Fly-flap.

from Benguella to the Cubangui, nor did I come upon any old trace of them. Their haunts were even at a considerable distance from the spot where we stood, as the hunters informed me they had still 6 days' march before them ere they could hope to fall in with the desired game.

A couple of days after my arrival, I received a visit

from the Sova of Cangamba, by name Moene Cahenga, who brought with him as a present four chickens and a large basketful of massango.

He was wearing the uniform I sent him, to which he had added a belt hung round with leopard-skins. He carried in his hand an instrument formed of antelopes' tails, with which he kept off the flies.

Field operations in the country appeared to be carried on by both men and women, who cultivate in small plots massango, cotton, a little manioc and, in even less quantity, sweet potatoes.

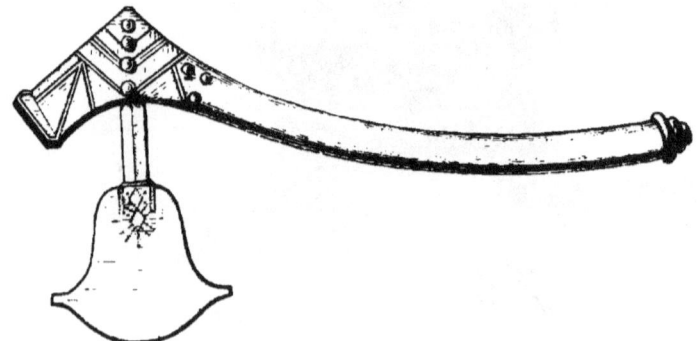

Fig. 63.—(CHIMBENZENGUE.) HATCHET OF THE AMBUELLAS OF CANGAMBA.

The natives work a good deal in iron, which they extract from the mines situated on the right bank of the river, to the north of Cangamba, and which we had passed on our way thither.

In Cangamba they reverse the practice common among the other Ganguella natives, as in that village the men make the baskets and the women the mats.

They weave the cotton they grow in rude looms, and produce cloths about the size of an ordinary towel, and very good they are.

Among the articles offered for sale was a little tobacco, which they asserted to have been cultivated in the country, but I did not see any growing in the plantations I visited.

Their arms consist of bows and arrows, and small hatchets.

The Cubangui, as it flows near Cangamba, is 16 yards wide and 19 feet deep, with a current running at the rate of 13 yards a minute.

The natives told me it contained plenty of fish, which I afterwards found to be the case; but all I saw in the neighbourhood were dried, and measured from 16 to 20 inches in length.

One soon learns to be thankful for small mercies, and manioc and dried fish appeared in our eyes

Fig. 64.—Ambuella Pipe.

material for a luxurious banquet, after being so long condemned to that abominable massango!

The river Cubangui is no exception to the general law of the streams on the African continent, being tolerably rich in crocodiles. They do not seem, however, to be of a very voracious kind, if I can believe the Ambuellas, who assured me there had been no instance of a human being falling a victim to their huge jaws.

I paid a farewell visit to the Sova, who was not by any means a bad fellow. As his people no longer offered for sale anything but massango, I begged him to favour me with some manioc and sweet potatoes.

He did so with a good grace, but the quantity was very small; and, as he gave it me, he apologised for the scanty supply upon the unanswerable plea that he had no more.

And so matters stood for three more days! Three days wherein we regaled upon massango!

Having obtained guides, a few carriers and a good store of the despised food, I decided upon making a fresh start on the 22nd July in the direction of the villages under the sway of the Sova Cahu-heú-úe on the river Cuchibi, through which runs the road originally traversed by Silva Porto. The former part of that track I abandoned at the Cuanza, to pursue a more northerly direction.

My guides informed me that I should have to travel through a desert country for 8 whole days, and that I must consequently be well provided with provisions. My invalids had, by this time, considerably improved with the long rest and more abundant food; but they were far from being recovered, so that Moene-Cahenga supplied me with 10 men to assist in the carriage of the massango with which he furnished me.

The guides having assured me that for a couple of days we should have to stick to the river's bank, I took it into my head to descend the stream in my india-rubber boat. Having ordered it to be conveyed to the river, I broke up my camp, and, entrusting the command of the caravan to Verissimo, I embarked with two young niggers, my attendant Catraio and another little fellow, about 12 years of age, called Sinjamba, the son of a Biheno carrier, whom I had selected for his knowledge of the Ganguella tongue, and converted into my interpreter.

I confess that it was not without a certain trepidation that I pushed off from the bank into the middle of an unknown stream, with mere children for companions and a fragile canvas boat beneath me.

The river, which has its source 30 miles to the N., and is, as I have already mentioned, 16 yards across and 19 feet deep at Cangamba, widens out a little below that village, and shortly displays a breadth of 40 to 50 yards, and occasionally even more.

Its bottom, varying from 10 to 19 feet in depth, is covered with a fine white sand, which evidently rests upon a bed of mud, as the aquatic flora is something wonderful.

Many kinds of rushes and other aquatic plants take root in the prolific bed, shoot their leaves and stems, in constant motion with the current, through nearly 20 feet of water till they reach the surface, where they display their multi-coloured and elegantly shaped flowers. Ocasionally this wealth of vegetation will occupy the whole expanse of the river, and seem to bar the passage of any floating thing. At the outset I had some hesitation about venturing my boat upon this aquatic meadow, as I thought it betokened too shallow a depth of water for navigation; but when my sound constantly gave me 12 and then 20 feet of depth, I acquired more confidence, and steered boldly through the floating garden.

There were points, indeed, where we came to a deadlock. These were places where the current, owing to some peculiarity of the river-bed, was scarcely perceptible, and the vegetation was so thick that it was more like a virgin forest than the growth of aquatic plants.

I saw abundance of fish darting hither and thither through the watery mass, many of them being at least a couple of feet in length.

Flocks of geese fled at my approach, astonished, doubtless, at so unseemly an interruption as the visit of such a monster to regions hitherto sacred.

Thousands of birds chirped and fluttered among the reeds and canes which lined the banks; the weight of a dozen of them producing scarce an impression on the gigantic grass-stems.

Occasionally a brilliant kingfisher would be seen hovering motionless in the air, until at a given moment it would descend from its lofty observatory like an arrow from a bow, and carry off its glittering prey from the surface of the water.

The birds were not the sole inhabitants of the clustering rushes on the banks. A sudden commotion amid the green stems would attract my attention, and a rapid glance would discover a crocodile just disappearing beneath the waters. Or the splash of a heavy body in the stream would betray the presence of an otter, either alarmed at our approach or, like the kingfisher, intent upon his daily meal. The whole place was instinct with life, and death, as usual, was following quickly in its train.

The river, whose general direction is north and south, winds in the most capricious manner; to such an extent, indeed, as to quadruple the journey. The right bank is a vast marsh of very variable width, attaining in some places to a thousand yards. It yields, in drainage, a huge volume of water, which produces a perceptible influence upon the growth of the stream.

Some three miles below Cangamba I came upon a bevy of 18 women, who were standing on the bank and fishing up small fry by means of osier-baskets.

At one of the turns of the river I perceived three antelopes, of an unknown species, at least to me; but, just as I was in the act of letting fly at them, they leaped into the water and disappeared beneath its surface.

The circumstance caused me immense surprise, which was increased as I went further on, as I occasionally

Fig. 65.—THE QUICHORO.

came across several of these creatures, swimming and then rapidly diving, keeping their heads under water, so that only the tips of their horns were visible.

This strange animal, which I afterwards found an opportunity of shooting on the Cuchibi, and of whose habits I had by that time acquired some knowledge, is of sufficient interest to induce me for a moment to suspend my narrative, to say a few words concerning it.

It bears among the Bihenos the name of Quichôbo, and among the Ambuellas that of Buzi. Its size, when full grown, is that of a one-year-old steer. The colour of the hair is dark grey, from one quarter to half an inch long, and extremely smooth; the hair is shorter on the head, and a white stripe crosses the top of the nostrils. The length of the horns is about 2 feet, the section at the base being semicircular, with an almost rectilinear chord. This section is retained up to about three-fourths of their height, after which they become almost circular to the tips. The mean axis of the horns is straight, and they form a slight angle between them. They are twisted around the axis without losing their rectilinear shape, and terminate in a broad spiral.

The feet are furnished with long hoofs similar to those of a sheep, and are curved at the points.

This arrangement of its feet and its sedentary habits render this remarkable ruminant unfitted for running. Its life is therefore, in a great measure, passed in the water, it never straying far from the river-banks, on to which it crawls for pasture, and then chiefly in the night-time. It sleeps and reposes in the water.

Its diving powers are equal, if not superior, to those of the hippopotamus. During sleep it comes near to the surface of the water, so as to show half its horns above it.

It is very timid by nature, and plunges to the

bottom of the river at the slightest symptom of danger.

It can easily be captured and killed, so that the natives hunt it successfully, turning to account its magnificent skin and feeding off its carcass, which is, however, but poor meat.

Upon leaving the water for pasture, its little skill in running allows the natives to take it alive; and it is not dangerous, even at bay, like most of the antelope tribe. The female, as well as the male, is furnished with horns.

There are many points of contact between the life of this strange ruminant and that of the hippopotamus, its near neighbour.

The rivers Cubangui, Cuchibi and the upper Cuando offer a refuge to thousands of Quichôbos, whilst they do not appear either in the lower Cuando or the Zambesi. I explain this fact by the greater ferocity of the crocodiles in the Zambesi and lower Cuando, which would make short work of so defenceless an animal if it ventured to show itself in their waters.

In an interview which I had at Pretoria with a celebrated antelope-hunter, Mr. Selous, I learned that he had heard my antelope spoken of by the natives of the upper Cafucue, a stream which, it appears, contained an animal similar to the one I had met with.

I regret that my very limited knowledge of zoology did not permit me to make a more minute study of a creature which I deem worthy the attention of men of science on account of the strangeness of its habits.

Resuming my narrative, I cannot but speak in the highest terms of praise of my mackintosh boat, which carried me so bravely over the waters of the Cubangui. Its only drawback was its restricted size, which confined me to so constrained a position that by 4 o'clock in the afternoon every joint in my body was aching.

I had seen no signs of my people since I left Cangamba, and, at the hour above mentioned, to the pain caused by my cramped posture were added considerable anxiety of mind and undoubted hunger of body. My young rowers were perfectly exhausted with fatigue. I made them pull up on the left bank, and ordered little Sinjamba to climb to the top of a tree in order that he might, from that elevation, see whether there were any signs on the other bank of the smoke of the encampment.

He thought that he perceived smoke in a N.W. direction, and consequently higher up the stream than the point we had then reached.

We therefore retraced our course, and, after some difficulty, I managed to get ashore upon the marsh on the right bank, and threaded my way towards the spot whence the smoke appeared to proceed.

I had walked about three-quarters of a mile, when I came upon traces of my caravan towards the south. The impressions of the men's footsteps might have misled me, but there was no mistaking the tracks of my goat and the dogs.

I returned to the boat, and again steered down the river. From time to time we pulled up, and the boy was set to climb a tree and look out, but the operation was repeated in vain.

Evening was now coming on, and my anxiety increased. Not only were we all desperately hungry, but I did not like sleeping away from the camp, on account of my chronometers, which would not be wound up.

The sun at last disappeared, and, as twilight is exceedingly short in these latitudes, I deemed it wiser to go ashore; which I did, with the two young niggers, on the left bank of the stream. Before we had settled ourselves down, I fancied I heard the distant report of a gun to the S.W. We at once got back into the boat, and pushed on vigorously upon hearing another report, to which I replied.

My signal was immediately answered by another, the flash of which I saw at some 200 yards' distance. I steered the boat in that direction, and shortly came upon my henchman Augusto, who was up to his waist in water in the marsh, along with a Biheno who had accompanied him. His delight at seeing me was very great, and he and his companion lost no time in pulling me out of the boat and conveying me across the marsh to the higher ground.

It was an arduous task, which it took half-an-hour to accomplish, but we reached terra firma at last. The lads, having secured the boat to some canes, quickly followed us. Augusto informed me that the camp was at some distance, and that we should have to cross a dense forest ere we reached it.

Unfortunately the night was pitch-dark, and locomotion was excessively difficult, owing to the unevenness of the ground and the resistance of the underwood.

Stumbling here, falling there, covering a dozen yards of ground in about as many minutes, tearing one's clothes, and one's flesh too, with the thorns of the brambles; such are the incidents which accompany a journey by night through a virgin forest.

After an hour of violent exercise, we heard, with indescribable pleasure, the report of rifles and the buzz of human voices.

They came from my own people, who were speedily gathered round us.

Verissimo Gonçalves appeared at the head of a troop of Bihenos, who insisted upon conveying me to the camp on a litter which they improvised with stout poles and the branches of trees.

It was in this guise that I returned to the encampment, where at midnight, beside a roaring fire, I appeased my hunger, made almost ravenous with a 36 hours' fast.

I remained in this spot the whole of the next day; but on the following one, at early morning, I commenced the passage of the river, which was a work of time, as my mackintosh boat was the only floating thing I had to trust to.

At about 9, I set out with my people along the left bank of the river, and an hour afterwards I· fell in with a brook, and started a good deal of game. Continuing on, I came to a halt at 1 o'clock, pitching my camp close to another little stream which, like the former one, is a tributary of the Cubangui.

While here, I was visited by two Ambuellas, who styled themselves "wax-hunters," and who informed my guides that it would be highly imprudent to proceed just then to the Cuchibi, inasmuch as a chief of one of the districts we should have to pass through had recently died, and that we should run the risk of being fleeced and maltreated, in accordance with the customs practised on such occasions.

The guides duly communicated this sinister intelligence to me; but, as I had resolved to go on, I told them I should do so in spite of the decease of all the *Sovetas* of the country. In proof of my sincerity, I broke up the camp next day, and, after a somewhat forced march of 6 hours' duration, I reached the right bank of the river Cuchibi.

I must here remark that several of my followers were on the sick-list, suffering from a malady which, though sufficiently painful, was not without a touch of the ridiculous. Some 18 or 20 of them had got a goitre!

CHAPTER VIII.

THE KING OF THE AMBUELLAS' DAUGHTERS.

The Cuchibi — The Sova Cahu-heú-úe—The Mucassequeres—Opudo and Capéu—Abundance—Kindness of the Aborigines—Peoples and Customs—A Ford of the Cuchibi—The River Chicului—Game—Wild Animals—The River Chalongo—An awful day—The Sources of the Ninda—The Tomb of Luiz Albino—The Plain of the Nhengo—Labour and Hunger—The Zambesi at last!

It was the 25th July that I camped on the right bank of the river Cuchibi.

The ground lying between this river and the Cubangui is clothed with a primeval forest, the vegetation of which is of the richest nature.

A botanist would there discover a vast field for lengthened study, so great is the variety of plants growing, one in the shadow of the other, in that enormous jungle.

In places it was most difficult to force a passage, and again and again the hatchets were drawn from the stout leather belts, to cut a path through underwood which had never probably before been invaded by the presence of man.

While traversing the forest, I became conscious of a most delicious and delicate odour, which I found to emanate from the flower of a tree that grew abundantly about me. There is not perhaps any known flower that has a more fragrant perfume than the blossom of the Oúco, for by that name do the natives designate the plant.

The configuration of the tree, the arrangement of its leaves, the flowers in clusters, and, above all, my ignorance of botany, induced me to speak of it in my diary as an acacia.

Some time after my return home, the apothecary of my village called upon me, and, turning over one of my sketch-books, he came upon a drawing of this particular tree. With the frankness which belongs to a villager, he observed: "Your Worship has committed a great blunder here. This can't be the flower of an

Fig. 66.—The Ot́co.

Flower ten times the natural size. The flowers form bunches 3 cent. long by 15ᵐᵐ in diameter. White petals, brown ovary and stamens; delicious perfume.

acacia, for it has only 2 petals and 3 stamens, whilst the blossom of the acacia has 5 petals and 10 stamens; it therefore belongs to the Papilionaceous family, and comes under the class of Leguminous plants, as I will show you presently, from the pages of Candolle..." "Don't trouble yourself," I said, as he was about to run off for his authority; "I will take your word for it. The flower is correctly represented, though I may be wrong about its parentage."

This tree, whose delicious flowers many a lady in

Europe would have rejoiced to possess, I never met with before reaching this particular spot, and I looked for it in vain as I approached the river Ninda.

There was another tree I also found in this forest that attracted my attention,—not, however, this time on account of the scent of its blossom, but the delicate flavour of its fruit,—and which the natives called Opumbulume. The fruit is, in shape, like that of the

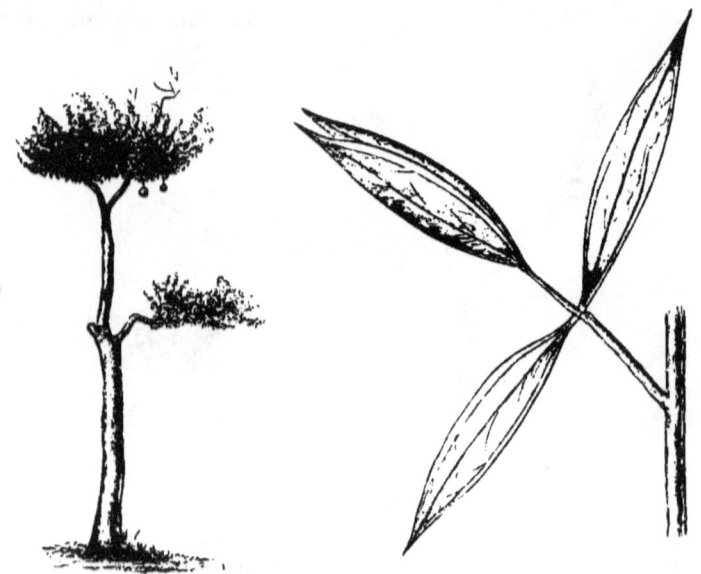

Fig. 67.—The Opumbulume.

Mapoleque, but with a different taste, and springs from a tree of a very dissimilar kind.

The river Cuchibi presents a different aspect from that of the other affluents of the Cuando, at least up to the point I investigated it. It flows through a long valley, enclosed by the gentle slopes of mountains covered with thick wood. This valley is perfectly dry, and not marshy, like almost all those through which flow the numerous streams of South-Western Africa, and which occasionally present a surface of water 6 miles in breadth. The river winds along, not in curves of short radius, like the Cubangui, but in a long undulated line, so that at a distance it seems almost straight.

Rich and abundant grasses cover its banks, but stop at the rather steep sides which enclose the river-bed, while the water, of the clearest crystal, courses along and allows the white sand beneath to be distinctly visible. It is entirely wanting in aquatic flora, so abundant in the Cubangui, although its fauna, of which I shall have something to say by and by, is by no means inferior.

Game was not wanting, and I was fortunate enough to bring down a *songue*, an antelope common enough on the banks of the Cuando and its affluents.

On that day several of my carriers complained to me of certain tumours which had broken out in the joints of the legs, and prevented them walking. Happily, the consumption of stores left a few of my men with nothing to carry, so that they were enabled to relieve the sufferers of their loads.

Most of the carriers were suffering from wounds of the ankle, instep and *tendo Achillis*, which there were no means of curing. I was soon at the end of my medical science borrowed from Chernoviz, and the same was the case with Dr. Chacaiombe, though his medicaments were supported by the most potent charms and stupendous arts of sorcery. Nothing would do them any good.

I attributed the sores, rightly or wrongly, to two causes, viz. the constant exercise of walking and the insufficiency or unwholesomeness of their food.

Let not my readers imagine that I am about to indulge in another tirade against that innocent massango. Oh, no! I am far too loyal an enemy to attack that pet aversion of mine during its absence:

I leave the massango alone, with the remark that it is in itself an inoffensive, and may be even a good and wholesome diet,—for those who take to it.

The food to which I refer, and to whose charge I lay in great part the fruitlessness of my efforts and those of Dr. Chacaiombe, is a very different one.

I have already had occasion to mention that the Bihenos will eat any mortal thing, and prefer their meat, when they can get it, in a state of putrefaction.

The circumstance which I am about to record will

Fig. 68.—RAT.

speak volumes upon this subject, and the fact that it does so must serve as my excuse for the somewhat disgusting narration.

My favourite hound, Traviata, had a litter of eight pups, all born dead. I gave orders to Augusto to bury them secretly in as inaccessible a place as possible, so as to remove them out of the way of the voracious jaws of my Bihenos; but two of those in the rear tracked out their burial-place, dug them up, and incontinently feasted upon them.

But it was not surprising that they should consider

young pups a delicacy, when they hunted out and devoured termites with as insatiable an appetite as an ant-eater, gathering them as they ran, and cramming them by handfuls into their mouths!

Rats, too, were a favourite food; one kind in particular, a small species which burrowed in the bee-holes, and doubtless fed on the honey, being highly esteemed by these epicures!

To return to my narrative. The part of the river Cuchibi where I was encamped was entirely unpopulated, and the guides informed me that it would take 4 days' march to arrive at human habitations.

Next morning we recommenced our journey, following the downward course of the river by the right bank.

About noon I discovered that many of my people were absent. I called a halt, and retraced my steps to look for them, when I found several of the fellows in the wood, bartering my cartridges, which they had stolen, with sundry Ambuella natives, for Quichôbo flesh, fish, and other articles.

On finding themselves discovered, they took to their heels, saving two, viz. the pombeiro Chaquiçonde and Doctor Chacaiombe, whom I caught in the act. The latter threw himself on his knees and prayed for pardon; but not so Chaquiçonde, who drew his hatchet and made a movement as if to strike me. I wrenched the weapon from his grasp, and gave him such a blow with the haft of it on the head, that it felled him senseless to the ground. I thought I had killed him: a mishap which occasioned my mind less pain than the cause which led to it, as it was the first time I had experienced positive insubordination from one of my own people. I turned to the men, who had now gathered about me, and ordered them to carry the wounded man into camp, which they at once did, the sight of the blood oozing from a rather ugly wound rendering them very silent and submissive.

On an examination of the hurt, I felt convinced that it was not mortal; and wounds in the head, if they do not kill at once, soon heal up. I did what my little skill dictated on behalf of the foolish fellow, and then called a council of the other pombeiros, to decide what punishment should be awarded for his double crime. The majority of them were for putting him to death, the rest for thrashing him within an inch of his life. As he had recovered his senses, I ordered him to be brought up for judgment, and having harangued him on the heinousness of his offences, ordered him to be set at liberty, with an injunction to "sin no more." My forbearance produced a great effect, though at first the fellows had a difficulty in believing that I was in earnest.

On the following day we had a march of 6 hours, still along the right bank of the river.

A good deal of game was visible in the course of the journey, but it was very wild, and the only animal killed was a *songue*.

This elegant creature differs considerably from the one on which the Bihenos bestow the same name between the coast and the Bihé country.

The one I shot measured 4 feet 7 inches to the shoulder, and was 4 feet 5 inches in length from the shoulder to the root of the tail.

Its short hair was of a reddish yellow, and uniform in tint. I found, on examination, that it could cover $17\frac{1}{2}$ feet in a leap, and I saw several of them go over the tops of canes which stood 6 feet out of the ground.

When brought to bay, it will fight with great courage and ferocity. The flesh is tasty enough, but, like that of all antelopes, it is very dry.

It feeds in herds, and always in the open; and sets

a watch while grazing. It takes to the forest only when it is closely pursued, on which occasions it will not hesitate to swim across a stream. Beyond the upper reaches of the river Ninda it disappears altogether.

Fig. 69.—THE SONGUE.

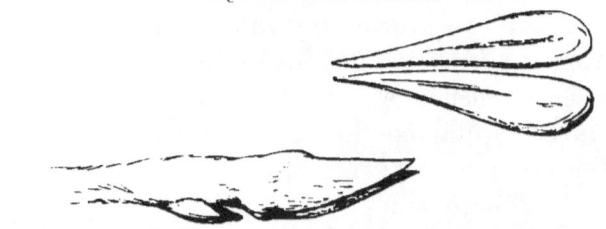

SLOT OF THE SONGUE.

Next day I pursued my journey. As I got lower down the stream, I observed that the level ground on each side kept increasing in width, and that the antelopes—the songues more especially—abounded.

Our stores of provisions had again run dry, and

on this day we consumed our last rations of massango.

It was on the 29th July, after a march of 3 hours, that we pitched our camp opposite the village of Cahu-heú-úe, where the Sova of the Cuchibi has his residence.

Before speaking of the Ambuellas tribe, and of the rich country watered by the Cuchibi, I wish to say a few words about my mode of travelling, or rather mode of life in Africa.

Undoubtedly all my predecessors have had their own particular system; those that come after me will have theirs, and each will think his own the best. My custom, therefore, with very rare exceptions, was the following. I rose at 5 o'clock; removed my clothes (as I always slept dressed and armed), and took a bath in water at a temperature of 65° Fahrenheit.

The English are accustomed to bathe in cold water, which is a capital tonic; I, for my part, used simply to wash for the purpose of cleanliness, and always had an iron pot with hot water ready, to produce the desired temperature. In referring to this subject, I must not fail to speak of my india-rubber bath, which came from the firm of Mackintosh of London. I found it a perfect treasure, and it still, after long and rough usage, is in capital condition. But this praise is due to all the india-rubber wares produced in England.

After my bath came my toilet. My wash-basin was formed out of a calabash, 18 inches in diameter. My towels were of the finest Guimarães linen.

The brushes, sponges, soaps and perfumery (I used a good deal of the latter in Africa) were of the very best quality, furnished by Charles Godfroy, whose goods, though very dear, are all excellent of their kind. My toilet over, in which I was assisted by my body-servant Catraio, that worthy collected and put carefully away all articles that had been used, and then brought me the chronometers, thermometers and barometer.

I wound up and compared the first, and registered the indications of the others.

By that time, young Pepeca had got tea ready, and brought it in.

It was served in a china tea-service to which I attached the very highest importance, it having been the gift of the wife of Lieutenant Rosa in Quillengues.

Fine as a sheet of paper, transparent in texture and elegant in form, that tea-service was my delight, and I thought the beverage never had the same flavour taken from any other vessel as out of that delicate porcelain cup.

Having swallowed three cupfuls of green tea without sugar, as I had not got any, the traps were packed up, and I gave orders to start; which, however, we rarely did before 8 o'clock, as it was next to impossible to get the men away from the fires, round which they gathered out of the intense cold.

Our order of march was the following. The lead was taken by Silva Porto's negro, Cahinga, bearing the flag, and immediately behind him came the cases containing the cartridges, and the wood and ropes for camp use. The other carriers followed indiscriminately, in single file, and I, Verissimo, and the pombeiros brought up the rear.

If a carrier, from any cause, had to fall out and lay down his load, the pombeiro under whose charge he was stopped also to assist and look after him.

During the journey I noted the course we took, and calculated our marches by the pedometer and watch. Our regular marches might be reckoned at from 8 to 10 geographic miles, although, when circumstances required it, they were much longer. Then came camping-time, and for the next hour all hands were employed in constructing huts.

In order to do so, some of the men were set to felling timber, others to lop off branches, and others again to gather grass. I meanwhile, if I had nothing else to do, stretched myself on the turf and slept, or at least tried to do so, till they came and told me my hut was ready.

It generally took about an hour; but, before I retired to my quarters, I used to take my observations for the meteorological record, which was regulated at 0 hr. 43 min. of Greenwich.

To learn the hour, I consulted a watch which Pereira de Mello had sent me from Benguella to the Bihé; it was in a brass case, was a pure cylinder, of Swiss make, with 8 rubies, &c., and went admirably.

At the proper time I called Catraio, who brought me the instruments. I used a swing-thermometer which had been the property of the ill-fated Baron de Barth; and each time I moved the instrument the whole of the Biheno carriers would stand at a distance watching the operation in wonder, and though it was regularly repeated every day, they always did the same thing, and always expressed the same mute surprise.

The observations having been duly registered, the young nigger, Moero, brought in the plates and my ration; for I cannot dignify with the name of *dinner* the handful of massango, boiled in water, which constituted the repast.

When it was over, if I were too tired to hunt up game or scour the neighbourhood, I employed my time in writing up my diary from my rough notes, in calculating observations, or in drawing. The ink which I used for all my work was obtained from small, so-called, "magic" ink-bottles, each of which lasted me from two to three months.

This system of taking notes during the marches and in the daytime, and subsequently transcribing them in the diary, gave me a duplicate record of my proceedings, and thus allowed me the chance of saving one, even if the other were lost. The daily notes were written in pencil, in little note-books, which, when full, I sealed up with wax. Besides the transcript of actual facts, I recorded in these little books all the initial observations, both astronomical and meteorological. On quitting Darban, I despatched them to Portugal viâ England, and they all arrived safely at Lisbon. They still remain there, unopened, whilst the copy which was made up from them remained constantly in my possession, and constitutes the basis of the narrative which I am now composing.

Until this journey was undertaken, I never knew the entire value of time, or how much can be done with it, if judiciously employed.

When night fell, the wood crackled on the temporary hearth, and gave me warmth and light. If I had no observations to make during the hours of darkness, or if —as was often the case—my fatigue compelled me to seek rest, I would lie down on the leopard-skins which formed my bed, using as a pillow the little valise in which I kept my papers.

A habit which I acquired during the journey, springing probably in the first instance from the cold which always preceded daybreak, was to wake regularly at 3 o'clock. I then rose and replenished the expiring fire, came to the door of the hut, just outside which hung a thermometer, and noted the point at which the mercury stood, for at that hour I could obtain a pretty correct minimum. I had not got with me any maximum and minimum thermometer, and therefore the figures which appear under these heads in my records are only approximate ones; the maximum temperature being that recorded at 1 hr. 30 min. of my register, or 0 hr. 43 min. Greenwich time.

From 3 A.M. till 5, my time was passed beside the fire, smoking; and I would often thus consume from 10 to 12 cigars, whilst thinking of my country and the dear ones I had left behind me.

How often at that hour—my time for meditation and sad reflections—did I not cogitate over present troubles and the uncertain future which lay before me!

At the time of which I am now writing I was on the Cuchibi, at 20 degrees E. of Greenwich and $14\tfrac{1}{2}$ to the S. of the Equator. I was far removed from all assistance of which I might stand in need, and where was I to seek for means and resources to pursue my onward journey?

From the Bihé to that point I possessed the few bales of cotton of which I made use; but the last pieces were now before me, and they constituted my entire stock of money.

In all the villages I passed through I met with more or less facility in bartering cotton-stuff for food, the *zuarte*, printed *zuarte* and white being most preferred.

On very rare occasions could business be done with the striped or trade cloth. Cowries, which are highly esteemed among the Quimbandes and disregarded by the Luchazes, recover all their value by the Cuchibi, although in the latter place they put them to a different use. Instead of employing them for the adornment of the head, they convert them into girdles, upon which they bestow extraordinary care.

The Maria II. beads have great value everywhere; but on the Cuchibi they are preferred to all other articles, powder only excepted.

On reaching the Cuchibi, I was asked, for the first time during my journey, for copper bracelets, and for wire to make them.

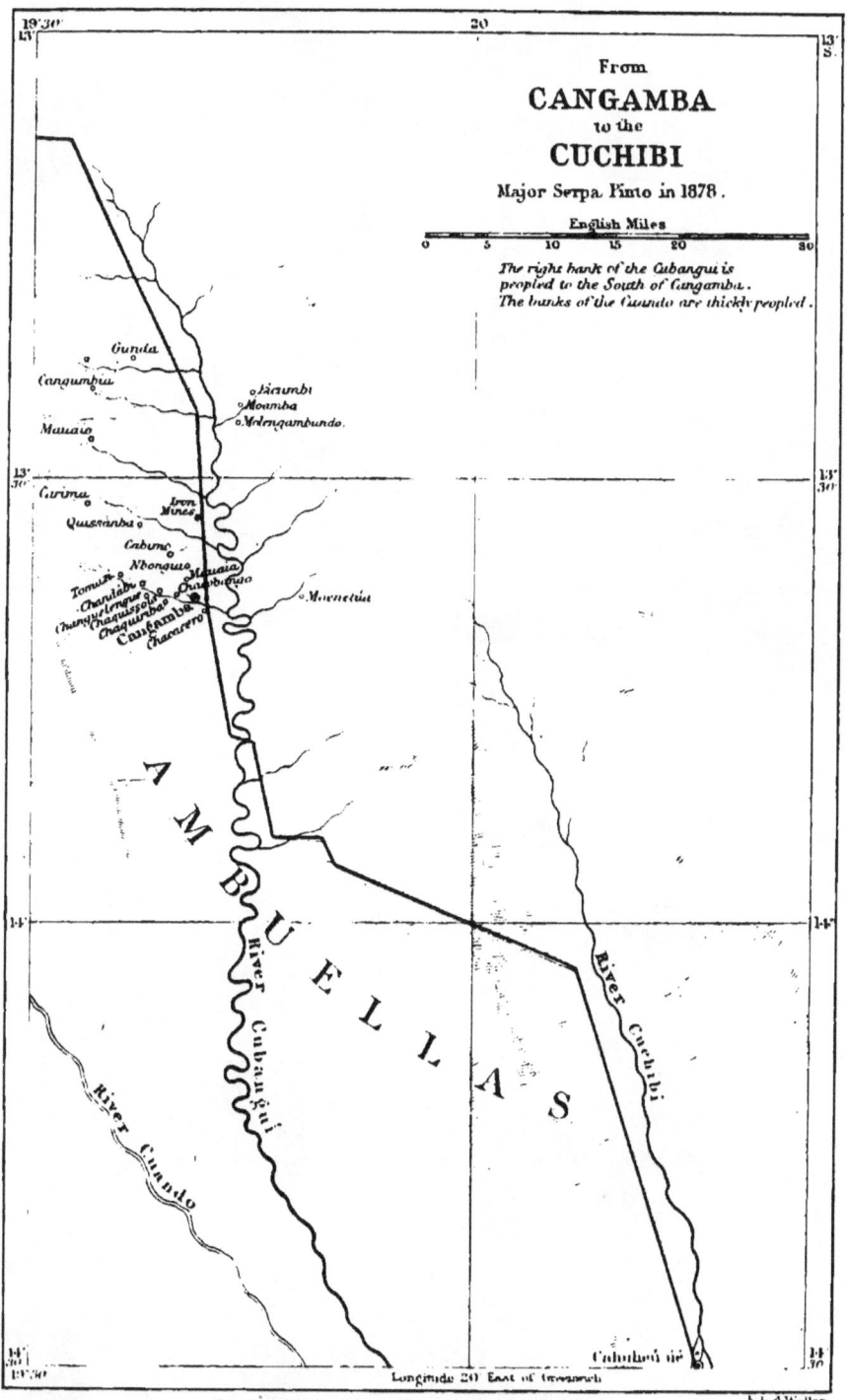

Immediately after my encampment was completed, a stranger came to me, stating that he was a Biheno, and had been left behind, on account of sickness, by a caravan three years before.

Being recognised by several of my carriers, I engaged him in my service.

I was now upon the track of the Bihé caravans, and as I intended remaining there a few days, I sent a little present to the Sova, and a message informing him of my determination.

I learned from the stranger Biheno that news had come of a revolution in the Baroze country, the native chief, Manuanino, having been expelled and another proclaimed in his stead, about whom little or nothing was known.

This intelligence was anything but agreeable to me, for I had heard that, though Manuanino was ferocious and sanguinary with his own people, he was very hospitable to strangers.

The Ambuellas among whom I was now sojourning were of the pure Ambuella race, whilst those of the Cubangui are a good deal mixed with the Luchazes.

The inhabitants of the Cuchibi are at enmity with the Ambuellas of the west, and they are frequently engaged in internecine war.

The Ambuella race occupy the whole of the country watered by the upper Cuando, and are collected more especially in the district where that river receives its confluents, the Queimbo, Cubangui, Cuchibi and Chicului.

The villages on the river Cubangui are constructed either on the islands which dot the stream or upon piles driven down into the river. As the inhabitants are the only people possessed of canoes, they repose at night in their aquatic habitations without the slightest fear of molestation.

The Sova lost no time in sending me provisions and a good supply of maize. What a treat was not that dish of boiled Indian corn! I saluted it with reverence, moved by the reflection that the reign of massango was for the moment at an end! His Majesty further sent me word that he would pay me a visit next day.

Early on the following morning I turned out for a stroll, but found walking difficult on account of the thorny nature of the underwood. Still, I managed to

Fig. 70.—The Sova Cahu-hec-ce.

get about 3 miles from the encampment, when I came across an enormous snare for catching game.

It was formed of a lofty hedge, which must have been a mile or two in extent, enclosing a nearly circular space. At about every 20 yards there was an opening in the fence, which led into smaller enclosures, carefully covered by a strong gin or urivi. A band of men being assembled, they beat the wood all round, and with loud cries frightened the hares, small antelopes and other animals, which, in their efforts to escape, darted into the enclosures referred to, and were caught in the urivi prepared for their reception.

On my way back to the huts I found in the wood an encampment of Mucassequeres, which gave evidence of being only recently abandoned.

The Sova called on me in due course. I found him a man somewhat advanced in years, of a sympathetic countenance, and rather a Jewish profile. He was extremely well dressed, wearing, over a sort of uniform, a cloak of white linen, with a large and handsome kerchief round his neck. His head was covered with a cap of red and black list, and in his hand he carried a concertina, out of which he wrung the most painful sounds.

He made me a fresh present of maize, manioc, beans and fowls, which I returned in the shape of a few charges of powder, the most valuable gift that could be made on the Cuchibi.

The old chief retired, extremely satisfied, and promised shortly to return.

In the course of our conversation, he informed me that the sovereigns of the Baroze were accustomed to demand tribute of him, and that, in order to avoid war, he had duly paid it, thus establishing a species of vassalage; that he knew little or nothing of the revolution on the Zambesi, and less of the new potentate who was in the ascendant, so that I still remained in utter ignorance of the state of the country I was about to enter.

During the afternoon my negroes captured in the forest two Mucassequeres, whom they at once brought before me.

The poor savages were trembling with fear, and gave themselves up for lost.

They knew a little of the Ambuella dialect, and by means of an interpreter we were able to understand each other. They imagined that sentence of death was about to be passed upon them, or that, at the least, the rest of their days was to be spent in slavery.

I desired my men to let them go, and return them their arms. I then told them that they were free, and might return to their people, and I gave them also a few strings of beads for their wives.

Their surprise knew no bounds, and they had much ado to believe that I was in earnest in what I said and did. Having ordered them something to eat, I inquired whether they would take me to see their camp.

After a warm discussion between them, carried on in a language unknown to all the bystanders, and completely different in intonation to any tongue I had hitherto heard spoken in Africa, they said they were quite willing to conduct me to their tribe, if I would trust myself to go alone. I accepted the offer, and immediately started with the two ill-favoured aborigines.

Accustomed as I was to the forest, I had much ado to keep up with my agile guides, who more than once had to wait for me to join them.

An hour's fatiguing walk brought us to a patch of cleared ground, in the middle of which was the encampment of the tribe.

Its inmates were three other men, seven women and five children.

A few branches of trees, bent downwards, with others interlaced in front, constituted their only shelter.

Of cooking-appliances there was not a semblance. Their food consisted of roots and fragments of flesh roasted upon wooden spits. Salt is quite unknown to them.

Both men and women barely cover their nakedness

with small monkey-skins. Their arms are bows and arrows.

I had come among them, but was perfectly at a loss how to act, now I had done so, for we neither of us could understand the other.

I thought the best thing to do was to ingratiate the women, so gave them a few strings of beads I had brought with me for the purpose. They received them, however, without the slightest sign of pleasure at the gift.

I was touched by the abject misery of these poor people. I examined them closely, and was much struck by their excessive ugliness. The eyes were small, and out of the right line; the cheek-bones very far apart and high; the nose flat to the face, and nostrils disproportionately wide. The hair was crisp and woolly, growing in separate patches, and thickest on the top of the head.

A few strips of the skin of some animal, encircling their wrists and ankles, constituted their sole ornament, and these were perhaps worn rather as amulets than for the purpose of adornment.

I managed to make my guides understand that I wanted to return, when, without leave-taking, they preceded me, and just as night fell left me at the edge of the wood, where I could hear the voices and merry songs of the people of my camp.

During my stay on the Cuchibi I managed to gather a few more scraps of information about these strange aborigines.

The Mucassequeres occupy, jointly with the Ambuellas, the territory lying between the Cubango and Cuando, the latter dwelling on the rivers and the former in the forests; in describing the two tribes, one may say that the latter are barbarians and the former downright savages.

They hold but little communication with each other, but, on the other hand, they do not break out into hostilities.

When pressed by hunger, the Mucassequeres will come over to the Ambuellas and procure food by the barter of ivory and wax.

Each tribe would seem to be independent, and not recognise any common chief. If they do not fight with their neighbours, they nevertheless quarrel among themselves; and the prisoners taken in these conflicts are sold as slaves to the Ambuellas, who subsequently dispose of them to the Bihé caravans.

The Mucassequeres may be styled the true savages of South tropical Africa. They construct no dwelling-houses or anything in the likeness of them. They are born under the shadow of a forest-tree, and so they are content to die.

They despise alike the rains which deluge the earth and the sun which burns it; and bear the rigours of the seasons with the same stoicism as the wild beasts.

In some respects they would seem to be even below the wild denizens of the jungle, for the lion and tiger have at least a cave or den in which they seek shelter, whilst the Mucassequeres have neither.

As they never cultivate the ground, implements of agriculture are entirely unknown among them; roots, honey, and the animals caught in the chase constitute their food, and each tribe devotes its entire time to hunting for roots, honey and game.

They rarely sleep to-day where they lay down yesterday. The arrow is their only weapon; but so dexterous are they in its use, that an animal sighted is as good as bagged. Even the elephant not unfrequently falls a prey to these dexterous hunters, whose arrows find every vulnerable point in his otherwise impervious hide.

The two races which inhabit this country are as different in personal appearance as they are in habits.

The Ambuella, for instance, is a black of the type of the Caucasian race; the Mucassequere is a white of the type of the Hottentot race, in all its hideousness.

Many of our sailors, browned by the sun and beaten by the winds of many a storm, are darker than the Mucassequeres, whose complexion besides has so much of dirty yellow in it as to make the ugliness more repulsive.

I regret exceedingly my inability to obtain more precise data concerning this curious race, which I consider to be worthy the special attention of anthropologists and ethnographers.

In my opinion this branch of the Ethiopic race may be classified in the group of the Hottentot division. In form it possesses many of the characteristics of the latter, and we may observe in this peculiar race a sensible variation in the colour of the skin. The *Bushmen* to the south of the Calaári are very fair of hue, and I have noticed some who were almost white. They are low of stature and thin of body, but exhibit all the characteristics of the Hottentot type. To the north of that same desert tract, more especially about the salt-lakes, there is another nomad race, that of the Massaruas, strongly built, of lofty stature, and of a deep black, who possess the same Hottentot type, and who indubitably belong to the same group. I was told on the Cuchibi that between the Cubango and the Cuando, but a good deal to the south, there existed another race, in every respect similar to the Mucassequeres, both in type and habits, but of a deep black colour.

In consideration, therefore, of the affinity of character, I have no hesitation in admitting that the Hottentot group of the Ethiopic race extends to the N. of the Cape as far as the country lying between the Cubango and the Cuando, passing through sundry modifications of colour and stature, due probably to the conditions under which they live, to altitude, to the great difference of latitude, or even to other causes that are less apparent.

The subdivisions of the Ethiopic race in tropical Africa will remain for a long time but indifferently known in Europe, on account of the difficulty of collecting reliable data wherewith to complete their study.

Where can we find any members of these barbarous tribes willing to allow their forms to be moulded? And, even if this difficulty could be overcome, how could the anthropologist convey thither materials to form his moulds, or how, if taken, could he convey his moulds to the coast? How could he manage to collect skeletons, or even skulls, in countries where the profanation of a grave might lead to the ruin of an expedition? How could he conceal from his own caravan, from the very carriers in his service, these human spoils, which would be regarded as articles of witchcraft?

Photography, of all means the most incomplete whereon to base serious studies, presents in itself almost insuperable difficulties.

In the first place, it is no easy matter to employ photography on a journey of exploration. Fancy, for instance, the conveyance of an apparatus, with its appliances in glass bottles, upon the head of a carrier who stumbles and falls at least a dozen times a day! My own experience will, I am sure, in this particular be supported by that of Capello and Ivens.

And, even supposing that that difficulty were got over, and that photography could be effectively employed, where is the native of the interior who would allow an apparatus to be set up, and stand before it as a subject for the camera obscura?

In the course of my narrative I shall have occasion

to relate an adventure which occurred to myself and a Swiss photographer, a M. Gross, where I managed to obtain a group of Betjuanos who were in a semi-civilised condition, after an expenditure of patience and time that was almost incalculable.

In respect of the Mucassequeres, I did not even succeed in making a satisfactory sketch with pencil and paper.

But to return to our narrative.

Fig. 71.—AMBUELLA WOMAN.

When my Mucassequere guides left me, as related, at nightfall at the edge of the forest, they uttered a few words, which probably meant a farewell, and disappeared in the darkness. The ruddy state of the atmosphere, due to the numerous camp-fires, and the sound of merry voices, guided my footsteps, and shortly after I found myself within the precincts of the encampment, where, to the notes of the barbarous music of the Ambuellas, the fellows were capering like madmen.

There were several Ambuella girls who were dancing with my carriers, and the bangles on their arms and wrists made a tinkling accompaniment to their motions.

I was much struck with the type of many of these girls, which was perfectly European, and I saw several

Fig. 72.—OPUDO.

whose forms, as they undulated in the dance, would have raised envy in the hearts of many European ladies, whom they equalled in beauty and surpassed in grace of motion.

What followed was calculated to increase my surprise.

It would appear that these Ambuellas, on the arrival in the country of a caravan, are accustomed to flock into the camp, to sing and dance; and, as night advances, the men retire, and leave their women-folks behind them. It is their hospitable custom thus to furnish the stranger wayfarers with a few hours of female society.

On the following morning, at daybreak, the visitors steal away to their villages, and rarely fail to return to bring gifts to their husbands of a night.

This custom led to an extraordinary adventure which befell myself.

Moene Cahu-heú-úe, the old Sova, sent me his two daughters, Opudo and Capéu.

Opudo was about 20, and Capéu counted some 16 years.

The elder was a plain girl enough, and was wonderfully haughty in manner; but the other was an attractive little creature, with a smiling and agreeable countenance.

From the moment of my setting foot in Africa I had determined to lead an austere life, a practice which gave me considerable influence over my negroes, who, seeing me only drink water, and detecting me in no *aventure galante*, looked upon me as altogether a superior being.

But now, notwithstanding my fixed determination, I was called upon to exercise no little restraint upon my feelings to resist the temptations of the younger daughter of the Sova Cahu-heú-úe.

Capéu only spoke the Ganguella dialect, which I did not understand, but Opudo talked Hambundo fluently.

"Why do you despise us?" she inquired in an imperious tone. "Are the women in your country more lovely and loving than my sister? Any way, we intend to sleep here; for it shall never be said that the daughters of the chief of the Ambuellas have been thrust out of his tent by a white man."

Here was a ridiculous position for a man to be placed

in! I was indeed so taken aback that I had not a word to say for myself.

Of course, a ready reply might have been found, but it was just the one that I had no intention to give.

There sat the two girls upon my leopard-skins, and there stood I. The large fire which separated us cast over the interior of the hut a ruddy light, somewhat subdued and softened by the green foliage which lined the cabin walls. The bright flame displayed to great

Fig. 73.—Capéu.

advantage the undraped figure of the young girl, whose languishing eyes were occasionally fixed upon me with an expression half-pouting, half-beseeching. My own looks wandered away, but involuntarily turned again and again to the statuesque and graceful figure.

Without, the noisy sounds of the barbarous music had ceased; the voices were more subdued, and silence was gradually taking the place of the previous uproar.

My braves were evidently selecting their companions

for the night; and there was I, still shut up with those irrepressible girls.

"We intend to remain here," repeated the haughty Ambuella princess. "I don't mean to expose my sister to the scorn of all the old women of the villages; and let me tell you, white man, that if you are a chief of the White King, I am the daughter of a Sova."

The ridicule of my position increased; I was compelled to put the firmest restraint upon myself, and, conscious that if I looked or spoke softly I was lost, I had to assume a severity of aspect and hardness of behaviour that were quite foreign to my character.

Still, things could not remain in the state in which they were, and I did not know how to alter them. I would have preferred, a thousand times over, risking a conflict with the warrior father to continuing this colloquy with the amorous little daughter.

Suddenly the skin which formed the door of my hut was raised, and some one entered.

It was little Mariana, who had overheard our limited conversation and came to the rescue.

She approached the fire, which she mended and replenished. Then, turning to the Ambuellas and repeatedly clapping her hands, as is the customary mode of complimentary salutation in the country, she uttered the words *Cô-qúe-tû Cô-qúe-tú*, and added: "The white man does not scorn you; but if he does not wish you to sleep here it is because I am the only one who does so, the white man is mine. My hut is alongside this one, and you are quite welcome to sleep there."

The daughters of Sova Cahu-heú-úe at once rose and left with Mariana, to whom I felt myself very greatly indebted for getting me out of my dilemma; but a few moments after, Opudo came back and whispered fiercely in my ear, "To-night we sleep elsewhere, but my sister does not mean to let you off."

I must confess it, this young woman inspired me with more fear than the wildest of wild cats could occasion.

I lay down on my couch, reflecting upon my extraordinary adventure, and beginning to credit, with more sincerity than I had hitherto done, the story of a certain Joseph who left his garment behind him in Egypt.

Next day the chief's daughters came in the usual way, to bring me presents. I gave them a few beads in return, and they retired without alluding to the scene of the previous night.

Shortly afterwards a messenger came from the father, to announce that he expected me that afternoon, and that he would send a boat to convey me to his village.

Our encampment had fresh visitors, in the shape of some cobras, which the negroes declared to be venomous, and several black scorpions, from 4 to $4\frac{1}{2}$ inches in length. One or two of the men were bitten by these disgusting reptiles, whose poison, however, produced no further mischief than violent pain and swelling of the parts affected.

The Ambuellas were the first people I fell in with on my journey who did not conceal their plantations in the forest.

Their fields under cultivation were all in the open, by the banks of the stream, and to this cause may be attributed their reputation as husbandmen.

The inundations which occasionally occur leave deposits on the land of the richest kind, and the fields become thereby naturally manured.

Although they do not, so to speak, irrigate the land —an operation which I never saw any African tribe practise—they nevertheless take the precaution, as I observed, of draining the ground by digging deep trenches beside their plantations.

My occupations had so engaged me during the day that it was not till evening that I remembered the canoe which the Sova told me would be in waiting near the river to convey me to his village.

On reaching the appointed spot my surprise was considerable at finding the frail skiff referred to *manned* by Opudo and Capéu, the two daughters of the chief! I do not consider myself a man of a particularly timid nature, but the sight of these two girls caused me some alarm.

This was no time, however, for indulging in such feelings, so I stepped into the canoe, and settling myself down, gave the signal for departure. The dexterity of these young women was remarkable, and they soon cleared the little creek or canal which led into the river.

The sun was fast nearing the horizon. The canoe

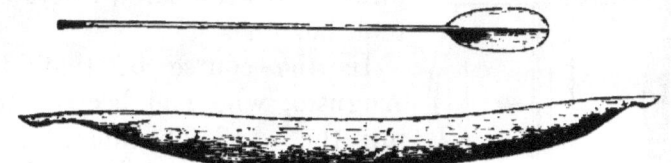

Fig. 74.—Cuchim Canoe and Paddle.

sped swiftly through the open spaces left by the abundant aquatic vegetation, which displayed upon the surface of the water a vast wealth of beautiful flowers. So thick were the clusters of Victoria-regias and many species of the Nenuphar, that at times they held us as in a net. On one occasion we were so imprisoned that I fully expected an upset, and in imagination saw those dark-skinned nymphs and myself struggling in the water among the crocodiles.

No such mishap, however, occurred. By a skilful manœuvre of the paddles we were set free, and Opudo then found her tongue.

"It is too late now," she said, "to go to our father's house. We waited for you long. We will return by land, and you shall come to-morrow."

Shortly after, at a convenient spot, we went ashore, and they accompanied me to the camp.

Night fell, and found the Sova's daughters again within my hut, conversing on indifferent subjects, whilst the sounds of dancing and merriment were heard without.

When the noise attendant on these festivities had ceased, they lay down near the entrance of the hut, beside the brightly burning fire. I wanted them to take up their quarters once more in the hut of little Mariana; but Opudo declined, saying she was a fawn of the forest, and little cared where she took her rest.

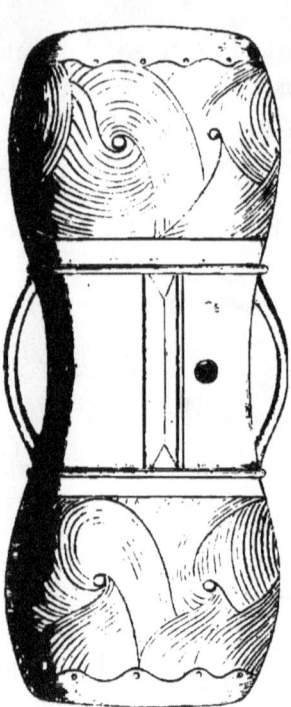

Fig. 75.—DRUM USED AT AMBUELLA FEASTS.

In the course of that day Augusto, who had been scouring the wood for game, fell in with a troop of small monkeys, the first I had come across in my journey from the coast westward.

On the following morning I paid my visit to the Sova; but, being desirous of avoiding further adventures, I got out my india-rubber boat, and proceeded to the village in that conveyance.

The canal I traversed communicated with an arm of the river, 22 yards wide by 19 feet deep, with a rapid current coursing along at the rate of 54 yards per minute.

The river divides, forming aits, little bays and marshes, which are the beds of thick and lofty canes. It is upon these small islands, themselves intersected

Fig. 76. — CAU-ET-HUE (TOWN ON THE CUCHIE).

by other channels, which form a perfect labyrinth, that these Ambuella villages are planted, springing from a marshy soil, on the level of the river. The houses are perfectly imbedded in the thick tufts of cane. Their walls are formed of reeds; their foundations are stakes driven into the muddy ground, and the roofs are composed of thatch.

As may readily be imagined, they are wretched habitations, badly constructed, and affording little effective shelter. Outside the doors, suspended from large poles, are immense calabashes, in which the inhabitants preserve their wax and other articles.

The huts themselves are filled with calabashes. Indeed, among the Ambuellas these useful vegetables perform the office of trunks, cupboards, and other household receptacles.

The store-houses only differ from the dwelling-houses in being raised upon stakes a couple of yards high, and therefore out of the reach of the inundations of the river.

On one of the small islands above referred to a little group of buildings constitutes the residence of the Sova Moene Cahu-heú-úe. One hut is occupied by himself, four more are assigned to his four wives, and the rest are store-houses.

I observed near the chief's own habitation a kind of rustic trophy, composed of the skulls and horns of animals and other spoils of the chase.

The Sova received me very graciously, he having two of his favourites by his side.

No sooner was I seated than my interpreter and one of the favourites commenced vigorously clapping the palms of their hands together, after which, scraping up a little earth, they rubbed it on the breast, and repeated many times, in a rapid way, the words *bamba* and *calunga*, terminating with another clapping of hands, not quite so vigorous as before. This completed the ceremony of introduction.

The chief expressed a wish to see my boat, and made a little excursion in it upon the river. His wonder at the floating power of this portable canoe knew no bounds; and again and again he urged upon me not to sell any such to the Ambuellas of the Cubangui, for that, if I did, he and his people were lost.

I pacified him on this head by the assurance that the whites did not wish for war between them, and would take all possible care not to furnish them with the means of waging it.

On our return to his island-home he sent for a calabash of *bingundo* and a tin cup, together with a pot of Lisbon marmalade, left there by some Biheno trader during one of his business journeys.

Having filled the cup, the chief allowed some drops of the foaming liquid to fall upon the ground, and, covering the place with damp earth, he drank off the contents without drawing breath.

The interpreter having informed him that I only drank water, he passed the calabash round to his favourites, who lost no time in disposing of what was left in it.

At noon I took my leave, and returned to the encampment.

I passed the rest of the day with a petty chief, the brother of the Sova, who informed me that he intended starting for the Zambesi by way of the Cuchibi and Cuando.

I found him to be a very intelligent fellow, speaking Portuguese pretty fluently, he having picked up the language while serving as a soldier in Loanda, to which place he had been sent as a slave when the horrid traffic was in the ascendant. He was a great hunter, and had frequently scoured the banks of the Cuando as far as Linianti during his sporting excursions.

THE KING OF THE AMBUELLAS' DAUGHTERS. 335

He assured me that the Cuando was completely navigable, that it was without rapids, and occasionally spread over so wide a bed as to present but little depth. Its aquatic vegetation was, however, so abundant and powerful that it not unfrequently barred the passage of any boats, and made navigation a matter of considerable difficulty.

He further asserted, and I had afterwards occasion

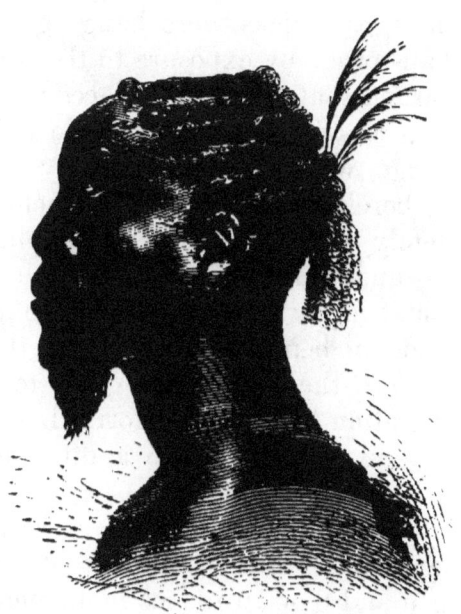

Fig. 77.—THE SOVA'S BROTHER.

to confirm the correctness of the assertion, that the river Cuando bears that name as far as Linianti, and thence to the Zambe either Cuando or Linianti, but never Chobe or Tchobe, as designated on the maps.

The Ambuella race continue on the Cuando the same system of existence as they practise on the Cuchibi, and the little islands are always selected for the establishment of their villages.

On the banks of the Cuchibi the preposterous head-dresses, which disappear among the Quimbandes, again become visible. Cowries, too, are once more in high esteem among the people, not for the purpose of adorning the head, but for the enrichment of their belts, which appear quite studded with them.

At the end of the canal where I embarked on my visit to the Sova I observed two faggots of thick sticks placed vertically at a few yards' distance from each other. From these sticks were hanging remnants of rush-mats, half-rotten by exposure to the weather. On inquiry, I learned that they were places where the rite of circumcision was practised upon male children of 6 to 7 years of age, who were subsequently turned adrift in the wood, bereft of their usual garment until they were completely cured, food being supplied them by those who had undergone the operation the year before. A piece of matting was given them to cover their nakedness, and, on being re-admitted to their village-homes, they left their mats hanging to the stakes where the operation had been performed.

I was also shown in this place another contrivance, of a very curious character.

Upon two stout pitch-forks, sticking half a yard or so out of the ground, was laid a sort of club, about a yard in length and 8 or 10 inches in diameter, wrapped tightly round with straw, looking for all the world like a large rolling-pin.

This notable apparatus was the work of a medicine-man of great fame, who had endowed it with most extraordinary virtues. When a husband had reason to suspect his wife of sterility, he sent for the doctor, who conducted her to this place of cure.

While muttering sundry cabalistic words, he passed the mysterious rolling-pin over her breast and sides, and so infallible was the result, as the Sova assured me, that nine months had barely elapsed than the desired end was attained.

Notwithstanding the deep faith reposed by the Ambuellas in this system of putting an end to sterility, I could not conscientiously recommend it for practice in Europe.

My relations with the aborigines continued to be most cordial and pleasant.

The Sova's daughters were indefatigable in bringing me presents, and, in fact, my own food and that of the young niggers about my person was supplied entirely by these good Samaritans.

Anything for which I expressed a wish was at once procured, and presumably their desire was to make others believe that closer ties than those of platonic friendship existed between us. I had learned by this time that they would have been held up to scorn if suspected of being repudiated by the stranger of their choice, and, out of regard for their feelings, I allowed them to have their own way.

We consequently lived on, the best friends in the world, and their co-operation was really of the highest importance in procuring me the carriers and stores of which I stood in need for traversing a vast depopulated space, where provisions would be simply unattainable.

By their exertions, chiefly, I was thus enabled to get together a good store of maize and a certain quantity of beans.

My pecuniary resources were drawing to an end, and, saving a quantity of powder in the shape of cartridges, a few beads, and a little copper for bangles, I had literally nothing left. Two of my carriers were bearers of the present I had reserved for the sovereign of the Baroze, the chief article being a small organ, having a couple of automatic dolls, which executed a dance to the sound of the music. This was a universal source of amusement to the aborigines. Augusto turned it to very profitable account, and many an egg did he conjure from the natives by the exhibition of the dancing figures. I was amused to find him testing his eggs by putting them into water, before accepting them in payment of the show, for, owing to the popularity of the entertainment, the eager sight-seers had more than once endeavoured to palm off on him eggs which they had surreptitiously abstracted from beneath a sitting hen!

Moene Cahu-heú-íc, no doubt upon the recommendation of his daughters, solved every difficulty as it arose, and actively aided me in my preparations for departure.

The daughters themselves had resolved to accompany me to the borders of their father's territory, and it was Opudo who assumed the command of my escort.

Before resuming the narrative of my journey, I deem it well to say a few words about the country of the Ambuellas and the people themselves, whose hospitality towards me had been so remarkable.

The Ambuella tongue is no other than that used by the Ganguellas, and which is heard for the first time to the east of the river Cuqueima.

Like the Hambundo, of which it is a dialect, it is exceedingly poor, very irregular in the verbs, and wanting in all those words which express noble and generous sentiments.

Can it be that these people are so unhappy that they do not feel the necessity of giving utterance to such sentiments in words, from the fact of their being foreign to their nature?

I tried in vain to discover if it were so, but I should have no difficulty in believing the conjecture a true one.

In this country, where I was received as a friend, and was therefore unbiassed by any influence adverse to the African, I sought in vain to read in the negro soul other than the most sordid cupidity, the most sensual appetites, cowardice in presence of the strong, and tyranny to the weak.

Of all the peoples I met with on my road, the Ambuellas were the greatest and most successful cultivators of the soil, which repays with wonderful prodigality the care and labour bestowed upon it.

Beans, pumpkins, sweet potatoes, ground-nuts, the castor-oil plant and cotton, are raised among enormous fields of maize of excellent quality. Manioc is likewise grown by these people, but I was able, unfortunately, to obtain little, owing to the destruction of the crops that year by inundations of an unusually heavy character.

Domestic poultry is the only live-stock possessed by the Ambuellas. Their mode of life, constantly disturbed by apprehensions of attacks from their neighbours, prevents them ever becoming herdsmen or shepherds; so that vast tracts of land, covered with admirable pasture, upon which enormous flocks and herds might be easily raised, are totally abandoned.

Cattle disappear with the last of the Quimbandes. Among the Luchazes one may meet occasionally with a few goats, and still fewer swine, whilst pigs abound in the Bihé and between the Bihé and the West Coast.

Why happens it that, in countries covered with the richest pasture, unvisited by the terrible tzee-tzee fly, and having all the requisite conditions for the breeding of cattle, no cattle whatsoever should be found?

An answer to this question may not perhaps be far to seek. Cattle constitute the greatest wealth of the African peoples, and, as a matter of course, always excite the cupidity of their neighbours; in fact, as I have already had occasion to explain, they are the permanent cause of the wars ever waging between the tribes residing between the West Coast and the Bihé.

This apprehension of being rich, and of being in consequence open to attack and robbery, no doubt has its weight in making cattle scarce between the Cuanza and the Zambesi. Among these barbarians, paradoxes are common enough, and we find principles planted and rooted among them that would with difficulty be comprehended in Europe.

The dog, that faithful and devoted friend of man, does not forfeit among the negroes his character as a sociable companion and trusty guard, and he is found among all the tribes of the Ganguella race. It is true that a variety of shaggy hounds and a few degenerate water-spaniels are almost the only specimens of the canine race that are met with in this part of Africa. Among the Quimbandes and Bihenos the dog is treated contumeliously enough—and little wonder, seeing that he is used there only as an article of food. He is most esteemed when he is dead, for his flesh is held to be a delicacy.

The Ambuellas, as I observed above, though furnished with the elements to become the first breeders of sheep and cattle in South Central Africa, possess neither cattle nor sheep, and breed common poultry only, and that of a small and inferior kind.

Among the inhabitants of the river Cuchibi there are no places set apart for the interment of the dead. Their Sovas are buried

Fig. 78.—Ambuella Hunter.

in any convenient spot in the wood, but the people find an unmarked grave in the mud by the riverside.

The customs of the Ambuellas may be designated as mild and sociable, and their hospitality, as will be gathered from what I have already recorded, is of the frankest order.

They are tolerable woodsmen, and gather a great deal of wax from the forest.

Women enjoy much more consideration among them than with any other tribes I had hitherto visited, for,

Fig. 79.—CHINGUÉNE.

One-fourth the natural size. Soft skin without scales. Brown back with darker spots; triangular shape, the back being the vertex; 3 belly fins, 2 subdorsal and 2 dorsal. Two muscular feelers upon the mouth, and two on the lower jaw. It belongs to a family that is very common in Africa and which comprises many species.

as a rule, they are the most abject slaves of their husbands.

The Ambuellas are skilful fishermen, which of course is not surprising, living as they do upon a river whose aquatic fauna is extremely varied.

In fact, of all the rivers I had come across, it was the richest in fish I had yet beheld.

During my stay I was able to obtain from the natives 18 different varieties, and they assured me that the specimens were far from complete.

I enumerate below, under the names furnished me by the aborigines, those which I was enabled to see and examine.

SMALL FISH, MEASURING UNDER 8 INCHES.

1. Mussozi	Skin fish.
2. Mango	do.
3. Chinguene	do.
4. Chibembe	do.
5. Limbumbo	do.
6. Dipa	Scaly fish.
7. Chitungulo	do.
8. Lincumba	do.
9. Nhele	do.
10. Lingumveno	do.

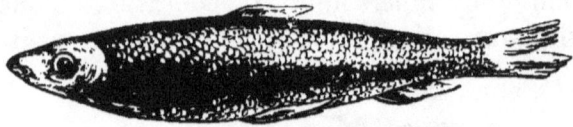

Fig. 80.—LINCUMBA.

Natural size. Scales broad and hard; grey back, silver-white belly; 5 belly-fins, 1 lumbar, all soft.

LARGE FISH, BETWEEN 8 AND 20 INCHES.

11. Chó	Skin fish.
12. Mucunga	Scaly fish.
13. Undo	do.
14. Chinganja	do.
15. Nassi	do.
16. Bula	do.
17. Ganzi	do.
18. Boci-io	do.

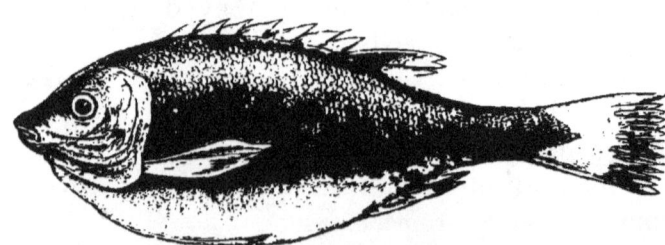

Fig. 81.—CHIPULO OR NHELE.

Natural size. Scales hard and small; back of a reddish-grey; belly reddish-white; 3 belly fins, 2 super-ventral, and 1 lumbar, occupying the whole back, and spiky.

Fig. 82.—THE CUCHIBI FORD.

SIX DIFFERENT LARGE MAMMIFERI INHABIT THE RIVER CUCHIBI.

1. The Hippopotamus.
2. The Quichôbo or Buzi (antelope).
3. The Nhundo (common Otter).
4. Libão (large Otter, spotted with white).
5. Chitoto (small Otter, perfectly black).
6. Dima (herbivorous, about the size of a small goat; without horns, existing under the same conditions as the Quichôbo or Buzi).

The reptiles, also, which inhabit the waters of the river are numerous, but the crocodiles are small and not of a very voracious character, and the cobras are not all of them venomous.

There is a great variety of the frog-tribe; but the Ambuellas do not specially distinguish them, but bestow upon all, generally, the name of Manjunda.

In canals and pools where the waters stagnate there exist myriads of leeches, as is the case with all the rivers in this part of Africa.

I had made a great provision of maize, and had got together carriers to convey it, under the command of the daughters of the Sova; so that on the 4th August I took my departure, after the most cordial adieux, and continued the descent of the river upon its right bank.

Two hours after leaving Cahu-heú-úe the guides pointed out to me a ford where the passage of the river might be safely effected. They themselves crossed over to show me the way, and I observed that a man of medium stature could wade breast-high for a space of 21 yards.

The river at that spot was between 77 and 87 yards in width. I stripped off my clothes, and proceeded to examine the ford. I found it was a narrow bar, with a depth immediately above and below it of 10 to 12 feet, with a very hard sandy bottom. The current of the river over the ford was at least 65 yards a minute. Under these circumstances the passage must be always difficult to a laden caravan.

I gave orders to commence the passage, which took a couple of hours in the performance. I remained the whole of that time in the water, with Verissimo and Augusto, the only two who were capable of swimming, ready to assist any of the men who should lose their footing. Not the slightest accident, however, occurred; nor, indeed, such was our care and precaution, was a single package wetted.

The passage of the river having been an excessively fatiguing operation, I determined to pitch our camp shortly after crossing, which was done on our arrival at the village of Lienzi.

The natives soon flocked in great numbers into the camp, bringing with them presents, and provisions for barter or sale. I never saw before in Africa so many fowls as were that day brought over by the Ambuellas. There was not a carrier or the youngest nigger but feasted that day on roast chicken.

I could not help being struck by the moderation and good-nature of the natives, which were really remarkable for an African people.

The whole of the men were armed with bows and arrows; a few of them carried assegais, and there were a good many who, besides the native arms, were possessed of long flint-lock guns of Belgian manufacture.

Both men and women cut the two front incisors in the shape of a triangle, but with a much more open angle than I observed among the Quimbandes.

Their arms are manufactured by themselves, the iron-work being of a very inferior kind. The iron itself is extracted from mines lying below the confluence of the Cuchibi and Cuando.

Those Ambuellas who use fire-arms greatly favour, as I have before had occasion to mention, the "Lazarinas" now manufactured in Belgium, and round the barrel of each gun they fasten a strip of skin of the animal it has brought down in the chase, which enables any one by a mere inspection of the weapon to know how many victims have fallen to its share.

The only result is to spoil the look of the gun and injure its utility by destroying the aim; but as they only risk a shot at 10 paces, they sometimes accidentally bring down their prey.

The hunter who had been most successful in his sport displayed no more than ten strips of skin about the barrel of his weapon.

This being the case, the poor people would get but few skins wherewith to cover their nakedness, if it were not for the snares they set in the woods.

Powder is an extremely rare commodity amongst them, and it is only very occasionally, with an interval

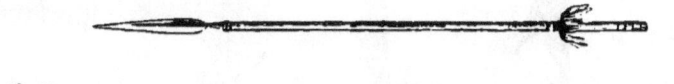

Fig. 83.—Assegais of the Ambuellas.

of months between, that a Biheno trader passes that way and sells them—for an enormous equivalent—the smallest possible quantity.

Among the Ambuellas who came into the encampment was one exceedingly pleasant-mannered fellow. He tried every possible means to convince me that I should be driving a capital bargain by exchanging a charge of powder for a fine cock he carried under his arm. I was much diverted with the ingratiating way in which he tried to persuade me to effect the exchange; and at last I told him that I would consent if he could kill the cock at 50 paces' distance with a bow and arrow.

He accepted the proposal, and I measured the distance.

The cock being set up at the allotted place, eight

arrows, each of which was infinitely wide of the mark, were fired at the intended victim.

A lot of the bystanders got quite excited with the sport, and at length a perfect cloud of arrows might be seen flying in the direction of the poor cock; but though the distance had been lessened to 40 paces, the best shot was still half a yard away from the mark. I then told the Bihenos that I would make the cock a present to whomsoever could kill it. The best marksmen from my caravan now came forward; the most successful of whom was Silva Porto's negro Jamba, who

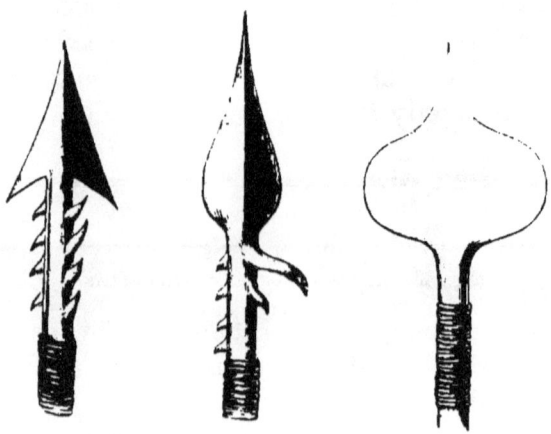

Fig. 84.—Ambuella Arrow-Heads.

planted an arrow within a quarter of an inch of the cock, which might, however, have lived and crowed for some time longer had I not put an end to the sport with a bullet from my Winchester rifle.

We discovered in the wood in which we were encamped an enormous quantity of white spiders, with bodies as large as the top of the thumb. They bit sharply, causing a violent though transitory pain.

Our camp was the resort of a considerable number of women, attracted probably by the presence of the daughters of their chief. They wore a great quantity of iron bangles round their wrists, about an eighth of an inch in thickness, of a quadrangular section, having the two outer edges indented. When they danced (and the Ambuellas are much given to dancing) the tinkle of these bangles had a very musical sound.

They compliment each other by repeatedly striking their open palms upon their naked breasts.

A custom which I met with among all the Ganguella people, but more rigorously observed in the Cuchibi than elsewhere, is noteworthy, and refers to the mode of addressing the Sova or Soveta.

The person who wishes to speak to the great man does not do so directly, but addresses his words to one of the negroes standing by the chief's side: he in turn repeats the remark or request to a second negro, who transmits it to the Sova. The reply passes through the same channel.

The explanation that was given to me of this arrangement was the following. The party who first speaks, by hearing his words repeated twice, has an opportunity of correcting any wrongful interpretation of his idea, and this is likewise the case with the party who answers.

My own notion, however, is that the Sovas have established the custom in order, during the triple repetition of the phrase, to secure time to prepare a suitable reply.

From Lienzi I went on a hunting-excursion down the river to its confluence with the Cuando, the position of which I marked roughly, being unable to take any observations; but I feel pretty confident in its being correct, on account of my having perfectly determined the position of Lienzi itself.

Close to the confluence of the Cuchibi I fell in with two large Ambuella villages, Linhonzi and Maramo, and between them and Lienzi a large Chimbambo village.

At the confluence of the river Queimbo is situated the village of Catiba, governed by a black from Cahuheú-úe, and subject to the Sova of the Cuchibi.

When I got back to the encampment I found my followers so given up to the delights of Capua, that there was no tearing them from the arms of the lovely daughters of this new African Nineveh.

The double intoxication produced by *bingundo* and love made the fellows deaf alike to entreaties and threats.

The Soveta of Lienzi came to call upon me, in company of a Mucassequere, his guest. I gladly engaged the latter to serve as my guide to the sources of the river Ninda, which I was desirous of reaching; and as the inclination was strong upon me to start at once, I called the pombeiros together and told them of my intention to go on with the Ambuellas and my young attendants, and that they might remain behind if they thought proper, but that, in any case, I should carry away with me the whole of the rations.

Having made them this communication, I set off under the guidance of the Mucassequere, and accompanied by the daughters of the Sova and their followers.

My Quimbares, seeing me in earnest, at once left the camp and followed me, leaving the Quimbundos and Verissimo's niggers behind.

After a painful march of 6 hours through the tangled forest, and where not a drop of water was met with, we reached the right bank of the Chicului, parched with thirst.

This river runs through a desert and swampy plain from 1800 to 2000 yards in width, and the forest, of unvarying density, only terminates where the marsh begins.

During the night the lions and leopards roamed incessantly around my encampment, roaring in the most frightful manner.

Next morning, at daybreak, I decided upon crossing to the opposite bank.

I passed the river at a place where a bridge had evidently at one time been thrown across the stream by Biheno caravans, and which I reconstructed. The passage was effected easily enough, but it was not so easy to reach the forest on the left bank, as we had to traverse the swampy plain, where we occasionally sunk to above our waists.

My little nigger Pepeca more than once remained with only his head out of the bog, and we had much ado to disinter him; and there were 1600 yards of this most trying and fatiguing swamp to get over.

The river I found to be 16 yards in width by 12 to 15 feet deep, with a current running at the rate of 45 to 50 yards per minute. I saw quantities of fish in the stream, both large and small, and a few crocodiles, but of no great size.

After crossing the river, I sighted, at about 600 yards down stream, a considerable herd of Songues, and, stealing a rapid march upon them through the brushwood, I managed to kill three.

My favourite goat Cora never left my side for a moment, and since she had heard the roaring of the lions was in a constant state of nervous alarm.

A good many birds were caught by my negroes, among which were a variety of quails and a white lapwing with white legs.

About 1 o'clock in the day my Quimbundos made their appearance, with the pombeiros, who in very humble guise entreated my pardon for not having come on with me the day before.

I was in no mood just then to be too hard upon them,

so forgave their temporary desertion, and shortly after I went on a fishing-excursion, with a very large net, by the aid of which I caught a good many fish, very similar to the mullet of the Portuguese rivers.

This same net or *barbal*, as it is called by the river Douro fishermen, was a present made me by my father, and which on various occasions proved our sole resource against the cravings of hunger.

The serious illness of one of my blacks induced me to remain a couple of days in that place, which put me out exceedingly; for, having with me a numerous company of Ambuellas, the provisions I had brought from the Cuchibi were disappearing rapidly, and I had before me an enormous tract of country to get over ere reaching the Zambesi, with the prospect of meeting with no resources beyond the spoils of the chase—a very problematical source of supply in Africa.

During one of those days the Ambuellas penetrated the forest in search of honey, guided by the *Indicators*, and were fortunate in securing a goodly quantity.

Many well-known naturalists from the time of Sparmann and Leveillant, the first who studied the habits of this curious bird, down to the most modern explorers, have made it the subject of lengthened description. Nevertheless, I must be pardoned if I say a few words more about so interesting a creature, dictated by my own experience and observation of its habits in Africa.

Whether the *indicator* is or is not a cuckoo is a matter which I will not attempt to discuss, but leave it to the authority of the Bocages and the Günters. Nor will I enter upon the other question, of deciding whether it should be called *Cuculus albirostris*, as Temminck asserts, or simply *Indicator* as averred by others. To attempt to describe it, with my limited knowledge of ornithology, would be presumption, so I shall confine myself to relating what I saw it do, and draw my own conclusions from the observation.

No sooner does man penetrate into one of the extensive forests of South Central Africa than the *indicator* makes its appearance, hopping from bough to bough, in close proximity to the adventurer, and endeavouring by its monotonous note to attract his attention. This end having been attained, it rises heavily upon the wing, and perches a little distance off, watching to see if it is followed.

If no attention be paid to it, it again returns, hopping and chirping as before, and, by its conduct and the manner of its flight, evidently invites the stranger to follow in its wake. The wayfarer yields at length, moved by the pertinacity of the bird, which, now flying, now hopping, but so as never to get out of sight of its follower, guides him through the intricacies of the forest, almost unerringly, to a bee's nest.

This is the most common instance, and the aborigines who are hunting after wax invariably allow themselves to be guided by its indications.

Some explorers, and among them the Portuguese Gamito, declare that the bird likewise entices men on to the den of the wild beast. This I cannot endorse of my own experience, as I have followed dozens of *indicators*, nor did I ever hear it affirmed by any native.

True, this restless bird has guided me and others to the carcass of some animal wasting in putrefaction, to an encampment recently abandoned, to a lake, or to other wayfarers; but why it should do any of these things is a mystery, inasmuch as it is in no wise a gainer by such a proceeding. But the fact remains that it shows man almost always the way to honey, and I believe it to be its fixed intention so to do; although, if the other destinations to which I have alluded, and which have produced the impression made upon many travellers, have been reached upon the road, it can scarcely be deemed remarkable in African forests.

For the same reason, it is very possible that a lion's den may stand in the way, without its being the bird's intention to entice the traveller into the beast's jaws.

Admitting, however, that the general rule, that the *indicator* points the road to where honey may be found, has exceptions, the examples of the rule being followed are so many and so various, that I have no hesitation in pronouncing this bird to be a friend to humanity.

I found near the river Chicului a cobra's skin, 22 feet long by 1 foot 5 inches wide, and was assured by the natives that even larger ones existed in the neighbourhood.

It was on the 9th of August that I was at length enabled to pursue my journey. I was very desirous that the daughters of the Sova of the Cuchibi should return home with their followers, as the rations we had with us were decreasing visibly, and my anxiety, as I surveyed the future, was anything but light.

After a march of 3 hours we fell in with a rivulet running S.S.E., and, having waded across it, we came upon a lake a couple of hundred yards wide, which we were also forced to wade through, with the water up to our waists.

The rivulet, which empties itself into the Chicului near its mouth, is the Chalongo, and is probably the same that figures on the maps under the name of Longo, and which, through erroneous information, our cartographers have made debouch into the Zambesi.

Whilst crossing the lake we observed several vultures hovering round, and descending to one particular spot about a quarter of a mile from us. Moved by curiosity, I went to see what was the special object of attraction of these disgusting birds; and, as I drew nearer, I saw a perfect flock of them whirling about some large carcass that was surrounded by hyenas, which made off without my being able to get a shot at them. On reaching the spot, I found an enormous *malanca* (*Hippotragus equinus*) recently killed by a lion.

The skin of this superb antelope was torn into strips by the lion's talons, and, what was remarkable

Fig. 85.—Malanca.

and inexplicable, the animal's hoofs were completely gnawed away.

The eyes had been torn out of the sockets, evidently by the rapacious birds.

My Quimbundos, who had followed in my steps, no sooner saw the malanca than they literally threw themselves upon it, and disputed with each other for the remnants of the carcass, mangled as it was by the beasts and birds of prey: an infinitely more horrible spectacle in my eyes than that which I had observed a few minutes before, when the wild beasts were at their dreadful work. Of the two, the men were the more deserving of the title.

And be it observed that at this particular time there was no necessity for their so acting, as I had killed game but recently, and the stores brought from the Cuchibi were not yet exhausted.

My very Quimbares could not resist the temptation, and soon joined the Quimbundos in their disgusting banquet.

Setting the caravan once more in order, we pursued our onward way, I pondering, as I went, on the power which savage life exercises over the negro.

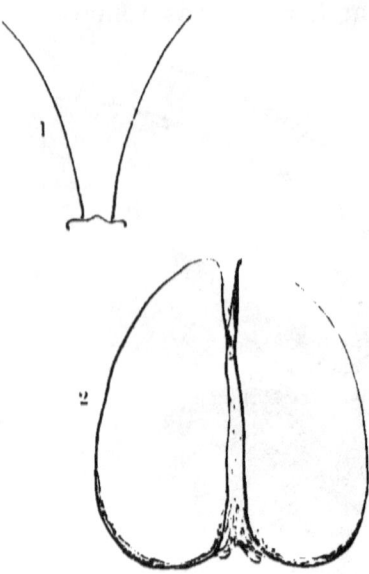

Fig. 86.—1. Direction of horns seen from the front.
2. Slot of the Malanca.

Here were these Quimbares, who came from Benguella, and were, so to speak, semi-civilised, and yet they were no better than the Quimbundos in savagery and brutishness.

I cannot at times help thinking that what is considered by many people in Europe as quite possible, viz. the civilising the negro in Africa is a pure chimera.

The civilising element is, at all events at the present time, so infinitesimal as compared with the savage element, that the latter must inevitably preponderate until the other shall assume far larger proportions.

In order to realise this dream of many exalted spirits in the old world, there must be a white man for every black upon the African soil, as by such means only can the element of civilisation be made to outweigh the savage.

We have an instance of this among the Boers of the Transvaal, who, European by origin, have in less than a century of time lost all the civilisation they brought with them from Europe, have become conquered by the savage element amid which they have been living, and now, though Europeans in colour and professing the faith of Christ, are the veriest barbarians in customs and behaviour.

It should be borne in mind that, in my journey hither, I had passed through many barbarous peoples, among whom not the slightest civilising element had ever permeated, and that, with the sole exception of the Bihenos, I had met with none in contact with the civilisation of the Western coast.

As I trudged on, I frequently thought of these things, and revolved in my mind a phrase which had been often repeated to me by my friend Silva Porto: "Mark this,—the best of the Bihenos are incorrigible; impress this truth upon your memory, and you are safe in dealing with them."

It was after I became acquainted with the Hambundo dialect that I learned to value them at their true price.

Occasionally at night, when quiet in my hut, I overheard the snatches of talk uttered around me, and no one would believe what I *did* hear.

One night, in particular, the subject of conversation turned upon certain episodes of a war that had broken out in the Bihé a year previous against certain Bihenos who refused to recognise the authority of the Sova Quilemo, when, in the midst of roars of laughter and other signs of approbation from his listeners, one of the fellows told the following story.

He made, it appears, a couple of prisoners, a lad and a young girl, and as the latter annoyed him with her tears and cries, he bound her arms strongly together, and cut off one of her ears. As this did not tend to make the unhappy victim more quiet, he struck his hatchet into her breast, but was careful to give such a blow as should not at once destroy life. The wretch then described to his auditors, in dramatic style, the contortions and groans of his poor captive, and narrated in sickening detail the mode by which he at length produced her death. It was a grim satisfaction to me to hear that he repented of what he had done, as her family, unknowing of her fate, had come to offer, in ransom, three able-bodied slaves, by whose aid he might have set up in business.

It would be painful and unnecessary to multiply descriptions of such horrible scenes as the foregoing; but it may well be believed that no chief of bandits in Europe requires, to maintain discipline among his horde of miscreants, greater energy and firmness than the European in Africa requires to lead and keep in hand savages of such a nature.

My camp was pitched at the source of a little brook called Combule, which, at about a mile from its spring-head, after a westerly course, empties itself into the river Chicului; its waters at no time being strong enough to move a mill.

I succeeded in persuading the Sova's daughters to return to their father's roof; and they departed, after a very cordial leave-taking. Even Opudo deigned to entreat me to take Cuchibi on my return, and come and live among them; whilst Capéu made her supplication still more eloquent by a glance of her eye,— one of those women's glances which are so powerful when spontaneous, and not acquired in the school of coquetry.

It was not without regret that I saw those two faithful girls depart; the only examples, as they were, that I had met with in Africa of natives forming an actual friendship.

When they were gone, my Mucassequere guide came to me and said:

"I have passed my life upon the road you are about to take from here to the Limbai, and I therefore know the country well. Have your best rifle always ready to your hand, and be ever on the alert while in the jungle, for you will be many days surrounded by wild beasts. Above all, be careful of the buffaloes of the Ninda. Many a grave will you pass—and some, too, covering the white man—that contain victims of those ferocious beasts. I am your friend, for you never did me harm, but gave me powder and beads, and therefore I put you on your guard."

After the departure of the Ambuellas, I was alone with my own people, and discovered, not without some alarm, that there was an enormous reduction in the provisions.

On the following day we penetrated into an extensive thorny forest, through which we had literally to cut our road.

After a fatiguing march of 5 hours, the most difficult and painful I had yet had in the country, we pitched our camp at the source of the river Ninda, having left a great part of our wearing apparel on the brambles by the wayside. Half-an-hour after our arrival I must have cut a very ridiculous figure in the eyes of any one but a native, as I was covered with bits of court-plaster where the thorns had picked out pieces of my flesh.

I had then at length reached the birthplace of that

Ninda which was so renowned for the ferocity of the denizens of its banks. The lions which favoured it had not yet succeeded in devouring me; but I could not help thinking if they wished to do so they must make haste about it, or they would find only the miserable remnants left by thousands of insects who considered me fair prey.

As evening fell, a cloud of flies, so small that they were impossible of measurement, swooped down upon the encampment, and, whirling about in a mad dance, penetrated the nostrils, the mouth, the ears and eyes, till we were nearly wild with pain and annoyance.

The encampment was surrounded by strong palisades and enormous abattis, and every precaution was taken to protect ourselves against any possible attack of wild beasts.

I had a visitation of another kind, in the shape of a violent attack of fever, which did not, however, prevent me getting up more than once during the night, and turning out to learn why the dogs were barking.

All through the dark hours the lions roared about the camp, and towards morning a chorus of hyenas helped to complete the infernal uproar.

I will not hesitate to put here upon record,—for the benefit of those who, in the enthusiasm of a fearless heart, have built up pleasant illusions concerning the delights of a sylvan life,—that where that life is thickly sprinkled with wild beasts it is positively most unpleasant.

I remained where I was till the afternoon of the next day, in order to determine my position, and then moved my camp a mile further to the eastward.

Close to the spot where I took up my new quarters was the grave of a fellow-countryman, the trader Luiz Albino, who was there killed by a buffalo. Among my followers I had Luiz Albino's favourite negro, old Antonio de Pungo Andongo, the very man I converted into the Sova Mavanda's tailor.

Luiz Albino had left the Bihé with a large quantity of goods which he was carrying to the Zambesi to trade with, and pitched his camp on the very same spot where mine was then standing. He turned out to give chase to a buffalo, which he wounded in the leg,—a proof that he was no great sportsman, as it was not the place to hit the animal.

Seeing it fall, he came back to camp, and summoning old Antonio (who was *young* Antonio then), bade him call the men and go out to seek a buffalo he had mortally wounded.

The Bihenos, who push caution to a fault, declined the task, and Albino, calling them a set of cowards, started off with Antonio for sole companion. On reaching the wood, the buffalo, which like all wounded buffaloes was waiting its chance to avenge the blow it had received, staggered to its feet and rushed at him. Luiz Albino fired off in quick succession, but without taking aim, both barrels of his gun; they had no effect in stopping the animal, which drove its horns into the unfortunate man's body.

Antonio fired with better success, but too late to save his master, for the corpse of the huge beast toppled over on to the corpse of the white man.

A strong wooden stockade, enclosing a piece of ground, some 15 feet square, protects a rude timber cross, and reminds the wayfarer of the necessity of having his rifle prepared and his arm steady when sojourning in these regions.

I had now reached the first stage of my journey where elephants appear, and I therefore deemed it advisable to send out some men as scouts, but they returned without discovering anything but old tracks of them. I then took a stroll into the wood, but saw nothing at which I could get a shot.

I continued my journey next day, still keeping on the right bank of the Ninda, without anything of note disturbing us on our march.

On the 13th August I shifted my camp 10 miles to the eastward of the spot where I had been staying the day before. A vague apprehension was beginning to take possession of my mind. The provisions were rapidly melting away, and I was still at a long distance from any country where resources were attainable. I beat about the forest for game, but without any result, although I perceived recent evidence of its existence; I thought I even saw some in the distance, but too far to be within rifle shot.

The following morning, the 14th, I happened to be marching along at the head of the caravan, with no other companion than young Pepeca, when, on reaching the place where I resolved to come to a halt for the day, I perceived an enormous buffalo quietly grazing.

Sheltered by the wood, I was able to get close up to him, and let fly at about 35 yards, aiming at the shoulder-blade, as he stood right across me. The animal fell like a stone, to my great astonishment, because the point I aimed at, if attained, would have produced death, it is true, but not so suddenly as occurred on this occasion. My surprise was redoubled, on examining the beast, to find that the ball instead of hitting him where I intended, struck just 6 inches higher, cutting the vertebræ and producing instantaneous death by the solution of continuity of the spinal marrow.

This circumstance caused me very grave reflection, inasmuch as such a deviation of the ball might one day be the cause of my ruin. So that, no sooner was the

Fig. 86 A.—THE BUFFALO.

encampment got a little straight, than I began testing the rifle at 30 yards.

The vertical deviation observable in firing at the buffalo continued to show itself.

It was my Lepage rifle, of large calibre and steel balls.

Its trajectory being very curved, the gunsmith had calculated the last groove of the rise for 87 yards, and as I had not used the gun for a shorter distance, I had not become aware of the danger I ran in aiming at 20 to 30 yards. So it happened that at those distances, and when, on account of the rifling, I could ill discover the culminating point of the aim, the vertical deviation was constant.

I at once took measures to remedy the defect, and little by little managed to deepen the groove of the rise, until I obtained the greatest precision at the shorter distance.

This episode, which I registered in my diary and now describe here, although of no interest whatsoever to the majority of my readers, may be useful as a hint to those who follow me in Africa, a hint that may perhaps serve them in good stead.

The river Ninda runs through a plain, slightly rising to the eastward, and which I was assured extends southward all the way to the junction of the Cuando and Zambesi.

Up to the point where I was encamped the forest descended thickly to the very brink of the river, but from that spot onwards there are merely groups of trees, scattered here and there over the enormous plain.

The Oúco, before referred to, is there a grand tree, and so abundant is it, and so plentiful its blossom, that for hours and hours the wayfarer is living in an atmosphere of almost overpowering perfume.

Next day we had a 6 hours' march, and deviated

somewhat from the bank of the river, as the reeds and canes which lined it were an obstacle to our progress. We then encamped alongside a lake of good water, not far from the little village of Calombeu, an advanced post of the sovereign of the Baroze country.

The people would sell us nothing, and provisions were beginning to get scarce.

Not liking my position, and yet being unable to resume my march on the following day, on account of several of the men being on the sick-list, I moved my camp a mile further to the eastward, and continued to draw water from the lake, or rather marsh, for it partook more of the character of the latter.

I was now in the vast plain of the Nhengo, lying 3900 feet above the level of the sea, which extends eastwards to the Zambesi and southwards to the confluence with that river of the Cuando.

The ground, dry in appearance, is little better than a sponge, yielding slowly but surely to the pressure of the body, the water oozing up and filling the cavity thus made.

During the nights that I was forced to stop there, I lay down on a bed that was dry enough, formed of dry leaves and covered with skins, but I always woke up in a puddle.

My life at this particular time was one of constant torment, as I failed to procure during the dark hours that refreshing sleep which repairs the fatigues of the day and helps one to bear better the troubles and apprehensions of the mind.

The dearth of provisions to which we were fast hurrying, the difficulties presented by the country that lay before me, the state of my own health, which I felt was deeply shattered, and the unsatisfactory condition of my people, among whom symptoms of insurbordination had frequently shown themselves, affected my spirits to such a degree that I was in a constant state of ill-humour.

On the 16th August there came upon me a feeling of despair. I felt myself alone,—completely alone,—not a man of my whole crew seemed to have a scrap of energy left in him.

Besides the tangible difficulties which rose up before me, all the fellows created, or seemed desirous of creating, imaginary ones. I had to interfere in, and to decide the minutest questions,—pure matters of detail with which I ought never to have been bothered at all.

I do not mean that my followers absolutely shirked their work, or purposely worried me, but what they did was done without heart or brain; they would obey an order if given to them in precise terms, but were incapable of procuring from others a like obedience.

Verissimo was no coward, but he was timid, wanting in strength of will, and irresolute; in fact, had no power to ensure obedience to his commands. Besides this, from being connected with some of the pombeiros, he had little or no hold over them. The consequence was, I had not only to issue orders, but myself see that they were obeyed.

I transcribe a few lines from my diary at this period, which will show the state of mind through which I was then passing.

"This upset me, and put me in a very bad humour. Great Heaven! how much will, how much pertinacity, how much energy are required by the man who, standing alone, surrounded by difficulties, created as much by his own followers as by natural causes, strives to fulfil a mission such as mine! Alone as I am in the centre of Africa with a great duty to perform and the honour of my country's flag to sustain, how much do I not suffer! Shall I ever bring it through untarnished? Truly, in situations such as these one must be either an angel or a demon, and at times I cannot help thinking I play the double part!"

It was on the day I wrote the above entry that we were put upon rations, and maize was the only article we had left.

Seated at the door of my hut, as evening was falling, I was finishing my frugal meal, and listlessly watching my carriers, who were squatting about and eating in silence.

It seemed as if some profound sadness had fallen upon the camp, and cast a spell over the whole of its inmates.

Suddenly my dogs started up and ran towards the wood, barking furiously.

A stranger man, followed by a woman and two lads, came from the bush, and, paying no heed to the dogs, entered the encampment, and giving a rapid glance round, advanced and seated himself at my feet.

He was a negro, whose bits of rags scantily covered his nakedness. What had once been a mantle hung from his bare shoulders. On his head he wore what only a great stretch of the imagination could call a cap, and in his hand he carried a stout stick.

His weapons were borne by the lads who followed him.

The energetic physiognomy, keen eye, and decision of manner of the stranger immediately commanded my attention.

"Who are you?" I enquired, "and what do you want of me?"

He answered me in Hambundo: "I am Caiumbuca, and I have come to seek you."

On hearing the name of Caiumbuca I could not restrain my emotion.

I beheld before me the boldest of the Bihé traders. The name of Caiumbuca, the old pombeiro of Silva Porto, is known from the Nyangwe to Lake Ngami.

In Benguella Silva Porto said to me: "Seek out Caiumbuca; engage him in your service, and you will have the best assistant you can meet with in all South Central Africa."

On reaching the Bihé I sought him high and low, but none could give intelligence of him.

"He is gone into the interior, and nobody knows where." This was the unvarying answer to my inquiries.

It happened that Caiumbuca was on the Cuando, just below the confluence of the Cuchibi, when, hearing of my approach, he started across country, with the woman and two young niggers, to join me.

I had a talk with him for an hour, I even read him a letter which Silva Porto had given me in Benguella for him: I made him my proposals, and by nightfall, everything being settled, I called my carriers together, and presented him to them as my second in command.

On the 17th August I made a forced march of 6 hours' duration, for our provisions were at an end, and it was absolutely necessary to reach human dwellings.

I camped on the right bank of the river Nhengo, which is in fact the Ninda, after receiving from the north an affluent of considerable volume, the Loati.

The Nhengo is from 87 to 110 yards wide, by upwards of 12 feet in depth, with an almost imperceptible current. At times it looks like a long lake in which thousands of aquatic plants are growing. Both banks are clothed with trees, so thick and luxuriant that their vigorous branches occasionally meet across the river, intertwine and form a romantic shade.

This important affluent of the Zambesi runs through the immense plain of which I have already made mention, a plain so spongy and humid that it may be considered a veritable swamp. It is the resort of myriads of snails, which drag their spiral houses through and over the short and wiry grass.

A vast number of tortoises (*Emydes*) also find a home in that congenial territory, and a few palm-trees, the first I had seen since leaving Benguella, were likewise visible, their elegant heads waving with the passing wind.

My negroes made a good collection of the tortoises, which hunger induced them to devour with avidity, notwithstanding the disgusting smell which emanates from these peculiar creatures.

Caiumbuca having informed me that at a short distance from the encampment there were some native villages, I decided upon stopping where I was for another day, in order to obtain provisions.

Early next morning I sent off some of the men for the purpose, but the natives turned out to be so shy, that they fled at their approach, and would not even listen to them.

Our position was now sufficiently serious, as we had literally nothing to eat, and all attempts both at hunting and fishing yielded no result whatsoever.

A group of our fellows, headed by Augusto, came running into shelter, pursued by several lions, which only retired on hearing the noise of the encampment.

I held a conference with Caiumbuca, and we decided on making a long march next day, as far as certain villages, which he called Cacapa, and where he assured me we should be able to obtain food.

We set off again, therefore, on the 19th, having eaten our last ration on the morning of the 17th!

The march was kept up for 8 hours, and at the close we pitched our camp near a lake, having left the banks of the river in order to get nearer to the villages.

In spite of the fatigue of the journey and the weakness produced by hunger, I sent off a deputation to procure provisions, Caiumbuca himself being one of the party. At nightfall they returned, but empty handed.

They obtained absolutely nothing. And the natives not only refused to part with any stores, but showed a disposition to hostilities!

What was to be done? To attempt another march, weakened as we were by hunger, was to run the risk of fainting and dying by the way. I therefore called the pombeiros together, and pointed out to them the precarious circumstances of the caravan; but I found them so disheartened, that they had not the ghost of a counsel to offer me.

I then summoned some of the negroes who had been up to the villages, and questioned them as to the actual existence of stores among the inhabitants. On their answering me in the affirmative, I took an immediate resolution, and I bade the pombeiros encourage their men with the assurance that next morning they should have a good feed.

When alone with Caiumbuca, I informed him of the resolution I had taken to march on to the villages and procure provisions at any cost.

In pursuance of this determination, at daybreak of the 20th I again sent off Augusto with a few negroes to the villages, to request the people to sell me maize or manioc, and explain the circumstances under which we were placed.

The only reply my envoys obtained were insults and threatened blows.

Thereupon I collected all my people who were not completely prostrated by exhaustion, amounting to some eighty semi-valiant men.

I placed myself at their head, and at once attacked the chief's compound; but, after a skirmish with no casualties, the place surrendered at discretion.

I lost no time in repairing to the general stores, which were full of sweet potatoes; and took out the quantity required to appease my people's hunger, returning afterwards to the camp with the petty chief and a few other negro prisoners. I then gave them the value of the potatoes in beads and powder, and set them at liberty, after pointing out to them that in future it would be far better to act in a more hospitable spirit. They were astounded at my generosity, and promised to supply me with everything I needed directly I applied for it.

At 1.30 p.m. of that day, the sky being clear, saving a dark bar on the horizon, a hurricane swept down from the N., shifting subsequently to the S.W. Its focus was fortunately three-quarters of a mile to the west of us, where it tore up trees and destroyed everything on its passage.

Even in camp the wind was so powerful that we were compelled to lie down flat on the ground until its chief violence was spent.

The thermometer rose from 20 to 32 degrees, and the barometer fell from 667mm. to 663. This was the most rapid barometric oscillation that I observed in tropical Africa.

At 2.30 the wind calmed down as suddenly as it rose, leaving the atmosphere completely covered with a dense fog.

The villages, which lay at somewhat less than a mile to the south of where I was camped, are called Lutué; but Caiumbuca informed me that among the Bihenos they are known by the name of Cacápa, on account of their being so rich in sweet potatoes, which in the Hambundo dialect is called *ecípa*.

The inhabitants of these villages, like all the aborigines of the Nhengo plain, are of the Ganguella race, subjected, by force, to the Luinas, or Barôzes. They are a miserable and intractable set.

Towards evening, a troop of Luinas arrived at the camp. It appears they were scouring the country round, and, learning that it was my intention to come to a halt in the neighbourhood, they gave me a look up.

The band was commanded by three chiefs, the principal of whom was named Cicóta.

These chiefs were wonderfully civil, and offered me their services. On my requesting them to obtain provisions for me, they replied that they were themselves badly off in the way of food; but that on the following day they would accompany me to other villages, where resources were to be obtained. They offered to guide me to the residence of the King of the Lui, and said I should want for nothing on the road so soon as I reached the Luina villages, now only at a short distance from us.

My Luina visitors were of good presence, tall, and robust. An antelope's skin, nicely dressed, passed between the legs and was fastened to the leather belt in front and at the sides, and an ample mantle of skins completed the costume. All three chiefs had rifles, of large bore, of English manufacture. The men carried shields of an oval shape, measuring 4½ feet long, by 20 inches wide, and were armed with a sheaf of assegais for casting. The chest and arms were covered with amulets. The wrists were adorned with bracelets of copper, brass, and ivory, and below the knees were from three to five very fine brass bangles. Their heads were the most remarkable, not on account of their hair, which was cut short, but from the way in which they were adorned.

Fig. 87.—LUINA SHIELD.

That of the chief, Cicóta, for example, was covered with an enormous wig made out of a lion's mane. The others had plumes of multi-coloured feathers, completely shadowing their features.

During the night we were visited by numerous scorpions, and some of my men were bitten by them.

Fig. 88.—The Chief Cicóta.

The ground continued spongy and wet, which must render life in such a country a perfect torment.

The palm-trees appeared in greater abundance; and the termites presented, in their ingenious habitations, a new form and aspect.

On August 22nd I broke up the camp, and 5 hours later pitched it again, close to the village of Canhete, the first occupied by the Luina race. A dense fog prevailed during the morning.

We passed through woods composed of enormous trees, but without any jungle, so that locomotion was easy and pleasant.

No sooner were my huts raised, than, at Cicóta's instigation, many girls came into camp, bringing me poultry, manioc, massamballa and earthy-nuts.

During the whole of the afternoon presents continued to pour in, which I returned in the best way I could, so that before nightfall there was abundance of food!

Fig. 89.—Ant-hills of the Nhengo.

I asked for tobacco (of which, by the by, I had still a good store), and salt. Salt! which I had not tasted for many months past!

To this they answered that, much to their regret, they were unable to comply with my wishes, as tobacco and salt could neither be given nor sold without a special licence from the King.

Hear it, ye free-traders! There is a country in the heart of Africa where there are two articles of contraband! Fortunately there are no custom-houses yet.

I paid a visit to the village of Canhete. In the fields there tobacco and the sugar-cane were growing in the utmost luxuriance.

The houses were built of reeds, covered with thatch; their shape being sometimes semi-cylindrical, with a radius of a yard and a half, and at others oval, of no greater height than the former.

The store-houses or granaries are similar to those of the Ambuella villages, but of smaller dimensions.

The Luinas returned my visit, and treated me in the camp to a war-dance, a very picturesque performance, in which a masked figure played the part of buffoon.

When night had fallen, my negro Cainga, whom I had despatched two days previously to the King to inform him of my arrival in his country, returned in safety.

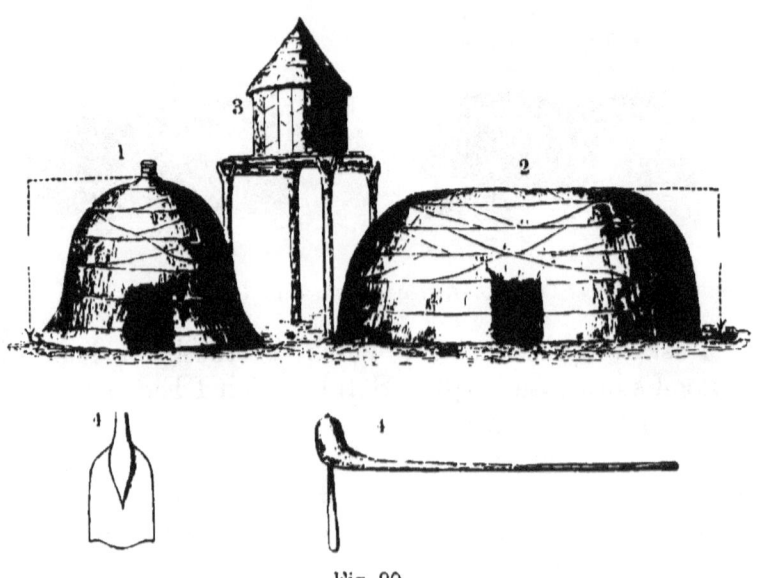

Fig. 90.

1 & 2. Luina houses, 4 ft. 7 in. high. 3. Granary. 4. 4. Luina hoe.

With him came various chiefs, bearing presents from his Majesty, among which were 6 oxen!

I could scarcely believe my eyes, and kept repeating, " Beef! We have really got beef to eat!"

Cainga told me that he seemed very proud at the idea of my visiting him by order of the Mueneputo, the White King, and that he intended giving me a splendid reception.

The news did not put me quite at my ease, for I knew the negroes well, and was aware of the treachery which frequently underlies their blandishments; nevertheless, I was not displeased at the intelligence.

With a view to display his greatness, he had ordered many boats to be got in readiness, so that my whole caravan might cross at the same time.

Cainga informed me that he was a young man of some twenty years of age, and that when he learned I was myself young, he said we should be friends.

I ate so much meat and so many potatoes, seasoned too with salt, which I obtained through contraband, that I made myself quite ill, and passed a horrible night.

The Luina chiefs, who came direct from His Majesty, brought orders for the people to supply me with what I wanted, gratis. This was a mercy, as I had little left to pay them with.

Just as I was breaking up my camp, fresh envoys arrived from the King, bringing salt and tobacco as a present, and with them a message desiring me not to follow the direct road to the mouth of the Nhengo, as he wished to punish the inhabitants of the villages lying on the route, by depriving them of the pleasure of my visit.

I sent word, in reply, that I intended to come by no other road, as it was the one that would suit me best. That I could not think of becoming the means whereby he should punish his delinquent subjects, and that if he did not send me boats at the point of the Zambesi I had indicated, I should cross the river without his assistance.

No sooner had we quitted Canhete than we fell in with a horrible swamp, which, though scarcely 550 yards wide, took us an hour to pass. We travelled eastward, and 3 hours later reached the village of Tapa, where I accepted a house offered me by the chief, it being impossible to camp without the precincts of the village, owing to the swampy character of the ground.

The houses in this place were shaped like a truncated cone, being built of canes plastered both inside and out with mud. The doors were about 1½ feet high and 16 inches wide. The house I occupied was surrounded by another one of granite, of concentric shape, with 3 feet greater radius. The roof covered both houses, and was also formed of canes covered with thatch.

The chief made me a present of a brace of fowls and some sweet potatoes.

I marked, at 2 miles to the south of my position, the large village of Aruchicho.

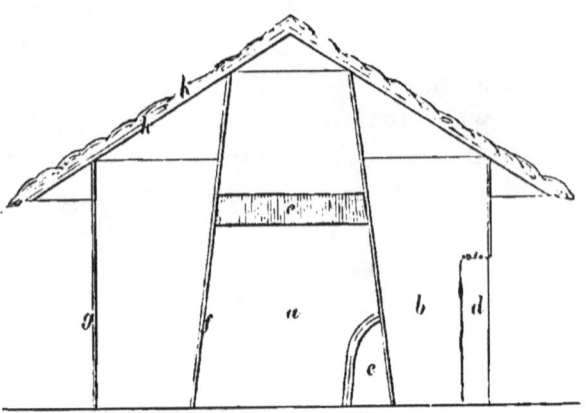

Fig. 91.—Vertical Section of a Luina House in the Village of Tapa.

a. Interior house. *b.* Space between the two walls. *c.* Internal door, 1½ ft. by 16 in. *d.* External door, 3 ft. by 1½ ft. *e.* Ventilator. *f.* Wall of cane and mud. *g.* Cane wall, 6 ft. *h.* Framing of cane. *k.* Thatch-roof.

On the 24th August we started at 8 o'clock in the morning. After crossing a swamp similar to that of the day before, we reached the right bank of the Nhengo at 9 o'clock; and, keeping along it until half-past 10, we arrived at that hour at the Zambesi.

With what enthusiasm did I not salute the grand river! A group of hippopotami were poking their huge snouts out of the water, at some 30 yards' distance, and two of them fell victims to their imprudence. An enormous crocodile, that was basking in the sun on an island hard by, shortly after shared the same fate.

I had thus appropriately saluted the mighty Liambai, by dyeing its waters with the blood of its ferocious denizens!

It was while the enthusiasm of my own people and of the numerous Luinas who accompanied me was at its height that the King's canoes arrived, and at mid-day we crossed to the left bank of the river.

Keeping still in an easterly direction, at 2 o'clock we fell in with another branch of the Liambai, which separates from it near Nariere. We therefore proceeded to a large island, on which there are hamlets, the chief of which is Liondo.

The branch of the river above referred to, although 164 yards wide, is very shallow, and we waded across it. On the other bank a good many natives were assembled, envoys of the King.

Still proceeding on, at 3 o'clock I arrived at a large lake near the village of Liara, which I crossed in a boat. This lake, formed by the overflowing of the Zambesi in the rainy season, is called Norôco.

My course continued easterly, and led through a perfect labyrinth of little lakes that had to be avoided, and it was not until 5 in the evening that I reached Lialui, the great capital of the Barôze or Kingdom of the Lui.

I found the King had drawn up a programme!

Two great surprises had, therefore, come upon me within scarcely more than as many days; for they were surprises to one who was already half a savage, and upon whose memory European customs were growing dim. Tobacco and salt were articles of contraband, and here was an African king making programmes!

Some twelve hundred warriors were drawn up in parallel lines, extending to the house I was provisionally to occupy, and one of the grandees of the Court, accompanied by 30 attendants, formed my suite.

On my arrival at the house, which had a large *pateo* or court-yard, surrounded by a cane fence, I found a daïs, on which I was compelled to sit to receive the compliments of the Court.

Four of the King's counsellors, with Gambella their President at their head, then arrived. At their back came all the grandees forming the Court of King Lobossi.

They seated themselves, and then began, both on their side and mine, a series of compliments and ceremonies, with a thousand protestations of friendship.

When they gravely retired, their place was taken by other envoys, who only left me when night had fallen.

I was then able to retire to the house set apart for me, and which was one of those semi-cylindrical ones I have already described; but I got little or no sleep, owing to my speculations on the future of my enterprise.

As the reader knows, I was at the end of my resources; and if the King did not energetically patronise my journey, what could I do? But for his generosity, I should not at that moment have had wherewith to stay the pangs of hunger.

He had informed me that next day we should meet and converse. What would be the result of our conference? That Gambella, the President of the Council, who had only recently left me, the man who, as I was informed on all sides, was the *de facto* king, how would he act towards me?

The following chapter will show that it was not without reason that an undefined presentiment of evil took possession of my mind, and caused me that sleepless night on the 24th of August, 1878.

END OF VOL. I.

www.ingramcontent.com/pod-product-compliance
Lightning Source LLC
Chambersburg PA
CBHW022103290426

44112CB00008B/532